Chemical Guide to the Internet

C. C. Lee

Government Institutes, Inc.
Rockville, Maryland

Government Institutes, Inc., 4 Research Place, Suite 200, Rockville, Maryland 20850

00 99 98 5 4 3 2

ISBN: 0-86587-519-7

Printed in the United States of America

Table of Contents

Preface v

About the Author vii

CHAPTER 1: Descriptions of Selected Organizations 1

(Alphabetical Order by Name of Organization)

CHAPTER 2: World Wide Web Resources by Subject 179

Chemical Analytical Instrumentation Manufacturers, 181
Chemical Analytical Methods, Services, and Information, 182
Chemical Computer Software Analysis, 183
CCA, 184
Chemical Information Service, 184
Chemical Professional and Trading Organizations, 186
Chemical Services, 187
Chemical Supplier/Manufacturer, 190
EAM, 195
Environmental Equipment Manufacturer/Supplier (Hardware), 195
Environmental Information Service, 195
Environmental Professional and Trading Organization, 197
Environmental Protection Agency, 197
Environmental Services, 200
Environmental Wastewater Information Service, 202
Environmental Wastewater Treatment Equipment Manufacturer/ Supplier, 203
Environmental Wastewater Treatment Facilities (Municipal), 203
Environmental Wastewater Treatment Services, 204
Government, 205
Life Cycle Analysis, 206
P2 Information Service, 206
PLI, 209
PLM, 209
PLS, 210

Recycling, 213
World Wide Web Engine, 214
World Wide Web Virtual Library, 214

CHAPTER 3: Academic Institutions 217

Chemical Engineering Programs, 219
Chemistry Programs, 234
Environmental Engineering Programs, 248

CHAPTER 4: Internet Discussion Lists 251

CHAPTER 5: Newsgroups 265

CHAPTER 6: Gopher Resources 275

Index 281

Preface

The objective of this guide is to help readers locate needed electronic information quickly and effectively. It focuses on the subject of chemicals and its associated fields.

The creation of the Internet has caused a revolution in our daily information exchange pattern. It is the most effective tool yet developed for data communication. The key to the effective use of the Internet system consists of three major components: 1) subjects of interest; 2) sources of the subjects; and 3) Internet addresses of the sources. This guide is based on these three components and covers comprehensive resources in the area of chemicals and its associated fields.

This book is divided into the following chapters: descriptions of selected organizations; World Wide Web resources by subject; academic institutions; Internet discussion lists; newsgroups, and Gopher resources. Entries in these chapters are drawn from six major areas.

Specifically, this book contains entries covering six major areas:

- Federal Agencies such as the U.S. Environmental Protection Agency and the Department of Health and Human Services that implement and enforce the laws and regulations on the subjects of chemically-related activities;
- Chemical industries which manufacture, process, and distribute chemicals;
- Green industries which promote pollution prevention or waste recycling to reduce the generation of chemical waste;
- Environmental industries which treat and dispose of chemical waste;
- Professional industries which provide meetings and exhibitions for information exchange;
- Institutions which provide education, research, and development of chemically-related activities.

This book is intended to be a tool to help readers search for needed data without wasting unnecessary time. It is a key to the secrecy of the Internet.

C. C. Lee

About the Author

Dr. C.C. Lee is a research program manager at the National Risk Management Research Laboratory of the U.S. Environmental Protection Agency in Cincinnati, Ohio. In addition, he is currently a member of the Policy Review Group to the Center for Clean Technology at the University of California, Los Angeles (UCLA). He is also the Chairman of the Sponsoring Committee to the International Congress on the Toxic Combustion Byproducts (ICTCB). He initiated the ICTCB and served as the Chairman of the First and Second Congresses which were held in 1989 and 1991.

Dr. Lee has more than 20 years of experience in conducting various engineering and research projects which often involve multi-environmental issues ranging from clean air and clean water control to solid waste disposal. He has been recognized as a worldwide expert in the thermal treatment of medical and hazardous wastes.

At the initiation of the U.S. State Department, he served as head of the U.S. delegation to the Conference on "National Focal Points for the Low- and Non-Waste Technology" (sponsored by the United Nations and held in Geneva, Switzerland). He has lectured on various issues regarding solid waste disposal in numerous and international conferences, and has published more than 100 papers and reports in various environmental areas.

Dr. Lee received a B.S. from the National Taiwan University, and a M.S. and Ph.D. from North Carolina State University. Before joining EPA in 1974, he was an assistant professor at North Carolina State University.

CHAPTER 1

DESCRIPTIONS OF SELECTED ORGANIZATIONS

(ALPHABETICAL ORDER BY NAME OF ORGANIZATION)

Chapter 1

Description of Selected Organizations

Abitibi Environmental Technologies

Description:	A developer of Supercritical Water Oxidation (SCWO) technology. The technology that is capable of eliminating 99.9999% of water-based organics from waste streams is ready for full-scale commercialization.
Internet via-web:	http://www.industrylink.com/aet
E-mail:	aet@inline.net
Phone:	(905) 822-4994
Fax:	(905) 823-9651
Mailing address:	2240 Speakman Dr., Mississauga, Ontario L5K 1A9

Aderco Fuel Additives

Description:	A manufacturer of industrial fuel additives for petroleum products. Founded in 1981, Aderco Chemical Products, Inc. with headquarters in Montreal, Canada, is one of world's leading suppliers of state-of-the-art fuel additives for large industrial consumers of petroleum products.
Internet via-web:	http://www.aderco.ca
Mailing address:	Montreal, Canada

Advance Scientific and Chemical, Inc.

Description:	A supplier of chemicals, laboratory glassware, and laboratory instruments. The company supplies (1) all types of chemical compounds for utilization by universities, corporations, and individuals; (2) all types and sizes of lab. glassware; and (3) all types of laboratory instrumentation.
Internet via-web:	http://www.sawgrass.com/advance
Phone:	(800) 524-2436
Mailing address:	2345 SW 34th St., Ft. Lauderdale, FL 33312

Advanced Chemical Design

Description: A manufacturer of borothene that is a non-CFC precision vapor degreasing solvent. Borothene can replace for 1,1,1 trichloroethane and cfc-113 and is a nonflammable, very high solvency alternative to 1,1,1 trichloroethane and chlorofluorocarbons. Based upon new halogenated hydrocarbon chemistry that has low toxicity and good cleaning power at a low cost. Borothene does not affect aluminum, magnesium, or ferrous metals as well as most plastics and elastomers. Works well in vapor degreasers and ultrasonic cleaners. This solvent power enables it to perform effectively and dissolve all greases, fats, oils, waxes, resins, gums, and rosin fluxes generally encountered in any metalworking, electronic, or precision cleaning application.

Internet via-web: http://www.coolworld.com/shopping/advanced/index.htm

Phone: (216) 382-2032

Fax: (216) 291-5949

Mailing address: 498 South Belvoir Blvd., South Euclid, OH 44121

Advanced Chemicals

Description: A manufacturer of chemicals for smoke and odor elimination. Other products include heavy duty, industrial strength, all purpose cleaner-degreaser with smoke and odor eliminator, glass and mirror cleaner, plus odor eliminator, lime eliminator, etc.

Internet via-web: http://www.chataqua.com/AC/

Advanced Visual Systems

Description: A supplier of scientific visualization. Their product items include AVS 5, AVS/express, toolmaster, an industry example, and sample images.

Internet via-web: http://www.avs.com

E-mail: webmaster@avs.com

Phone: (617) 890-4300

Fax: (617) 890-8287

Mailing address 300 Fifth Ave. Waltham, MA 02154

Aerated Biological Surfaces, Inc. (ABS)

Description: A manufacturer of municipal and industrial wastewater treatment devices. ABS manufactures systems capable of handling between 5,000 and 25,000 GPD of waste.

Internet via-web: http://execpc.com/~abs/
E-mail: abs@earth.execpc.com
Fax: (414) 261-7760

Aerojet Chemicals

Description: A manufacturer of various chemicals. Specific chemicals of interest can be found by searching through the Internet homepage, WWW Chemicals, at the Internet address, "http://www3.ios.com:80/~ilyak."

Internet via-web: http://www3.ios.com:80/~ilyak/ind030.html
Phone: (916) 355-5783
Fax (916) 355-4936
Mailing address: PO Box 13222, Sacramento, CA 95813-6000

Agency for Toxic Substances and Disease Registry (ATSDR)

Description: An agency of the Public Health Service in the U.S. Department of Health and Human Services. The mission of ATSDR is to prevent exposure and adverse human health effects and diminished quality of life associated with exposure to hazardous substances from waste sites, unplanned releases, and other sources of pollution present in the environment. ATSDR is directed by congressional mandate to perform specific functions concerning the effect on public health of hazardous substances in the environment. These functions include public health assessments of waste sites, health consultations concerning specific hazardous substances, health surveillance and registries, response to emergency releases of hazardous substances, applied research in support of public health assessments, information development and dissemination, and education and training concerning hazardous substances.

Internet via-web: http://atsdr1.atsdr.cdc.gov:8080
E-mail: Mike Perry/lmp1@atsoaa1.em.cdc.gov
Phone: (202) 690-6985
Mailing address 1600 Clifton Road, Atlanta, GA 30333

Agency for Toxic Substances and Disease Registry (ATSDR) Science Corner

Description: A guide (MENU) to search the World Wide Web for environmental health information. The primary focus is to find and share global information resources with the public on the linkage between human exposure to hazardous chemicals and adverse human health effects. New environmental health information resources are cited, searched, and documented as they come on-line.

Internet via-web: http://atsdr1.atsdr.cdc.gov:8080/cx.html

E-mail: chx@atsdr.em.cdc.gov

Mailing address 1600 Clifton Road, NE, MS, E-28, Atlanta, GA 30333

AGRA Earth and Environmental Limited

Description: A waste management company. AGRA Earth and Environmental is a consulting firm in Canada, the USA, and around the globe and has more than 1,200 employees who provide expertise in a wide variety of disciplines including environmental, geotechnical, and materials engineering. Its provides some useful solid waste-related technical information to the global community. It also links to other garbage-related sites on the Internet.

Internet via-web: http://www.kingsu.ab.ca/~agra/agra.html

E-mail: mkostec@freenet.calgary.ab.ca

Ajax Chemicals

Description: An Australian-owned manufacturer and distributor of a diversity of products. The products range from industrial and fine chemicals to laundry and pool care products.

Internet via-web: http://www.science.com.au/ajax/

Mailing address: Sydney, New Zealand

Akromold Inc.

Description: A plastics engineering company that offers tooling and molds for the plastics industry. The company is equipped with CAD/CAM and CNC automation technologies and is dedicated to providing a high quality mold delivered on time. Akromold features in-house tryout capability for final quality approval.

Internet via-web: http://www.polysort.com/akromold

Phone: (216) 929-3311

Fax: (216) 920-1338

Mailing address: 1100 Main Street, Cuyahoga Falls, OH 44221

Akzo Nobel

Description: A manufacturer of more than 7,500 specialty chemical compounds. The company also manufactures pulp and paper chemicals and a broad range of products including chemicals, coatings, fibers, healthcare, and pharmaceutical. Akzo Nobel is an international company, headquartered in Arnhem, the Netherlands, with locations throughout the world.

Internet via-web: http://www.akzonobel.com

Mailing address: Headquartered in Arnhem, the Netherlands

Alberta Research Council (ARC)

Description: Alberta is one of the Canadian provinces. ARC is a leading technology corporation committed to advancing the economy of Alberta through the application of science, engineering, and technology in six major strategic business areas: biotechnology, energy, environment, forestry, information, and manufacturing.

E-mail: techline@arc.ab.ca

Alitea USA

Description: A manufacturer of flow injection instruments. This analytical instrument can be applied to chemistry, biotechnology, biology, process control, environmental analysis, and sensor technology.

Internet via-web: http://www.flowinjection.com/flowinjection

E-mail: fialab@flowinjection.com

Phone: (206) 453-4135

Fax: (206) 454-9361

Address: PO Box 26, Medina, WA 98039

Aliweb

Description: An information search engine. Subjects that can be searched in this engine cover most human activities including arts, business, computers, education, engineering, entertainment, government, health, medicine, news, recreation, science, social science, society, and culture.

Internet via-web: http://www.traveller.com/aliweb/

Alliance for Environmental Technology (AET)

Description: An international alliance of chemical manufacturers and forest products companies dedicated to advancing the environmental performance of pulp and paper manufacturing. AET supports the use of Elemental Chlorine-Free (ECF) technology, based on chlorine dioxide.

Internet via-web: http://aet.org/index.html

AlliedSignal

Description: A manufacturer of more than 60 nylon and 80 polyester products for thousands of end-use applications in tires, mechanical rubber goods, broad wovens, narrow fabrics, and cordage.

Internet via-web: http://www.polysort.com/allied

Phone: (216) 668-4880

Fax: (216) 668-3807

Mailing address: 4040 Embassy Parkway, #150, Akron, OH 44333-1700

Alpha Analytical Labs

Description:	A full-service environmental analytical laboratory. The company has compiled a collection of links to Internet resources related to environmental science and regulatory affairs.
Internet via-web:	http://world.std.com/~alphalab/
E-mail:	info@alphalab
Phone:	(508) 898-9220
Fax:	(508) 898-9193
Mailing address:	8 Walkup Dr., Westboro, MA 01581

Alta Vista

Description:	An information search engine. Subjects that can be searched in this engine cover most human activities including arts, business, computers, education, engineering, entertainment, government, health, medicine, news, recreation, science, social science, society, and culture.
Internet via-web:	http://www.altavista.digital.com/

ALToptronic AB

Description:	A supplier of sensors for chemical analysis using semiconductor lasers and fiberoptics. The information includes laser diode spectrometer: LDS 3000, technical description, general specifications, gases, publications, etc.
Internet via-web:	http://www.altptronic.se/

Aluminum Industry World Wide Web Server

Description:	A uniform resource locator (URL) directory that hosts a variety of servers who are involved in the production of aluminum and aluminum products.
Internet via-web:	http://www.euro.net/concepts/industry.html
E-mail:	concepts@euronet.nl

American Chemical Society (ACS)

Description: A professional society that provides information on ACS products and services. The ACS is the world's largest scientific society and has a membership of nearly 150,000 chemists and chemical engineers. The Society is recognized as a world leader in fostering scientific education and research, and promoting public understanding of science.

Internet via-web: http://www.acs.org

E-mail: his92@acs.org

Phone: (800) ACS-5558

Mailing address: 1155 Sixteenth Street, NW, Washington DC 20036

American Chemicals Company, Inc.

Description: A chemical, solvent, oil, lubricant, and specialty product distributor. The company's products and services include commodities and solvents, chemical specialty products, oils and lubricants, research and development, survey, materials safety data sheet database, Chemical Week Magazine, and hot links in the chemical industry.

Internet via-web: http://uc.com/acci/

E-mail: american@panix.com

Phone: (201) 344-3604

Fax: (201) 578-4513

Mailing address: 49 Central Avenue, South Kearny, NJ, 07032

American Institute of Chemical Engineers (AIChE)

Description: A professional organization for chemical engineers. The organization promotes excellence in the development and practice of chemical engineering.

Internet via-web: http://www.che.ufl.edu/aiche/

E-mail: kirmse@che.ufl.edu

Mailing address: 345 East 47th Street, New York, NY

American Mold & Engineering

Description:	A plastics engineering company that designs and builds plastic injection, structural foam, die cast, and compression mold tooling.
Internet via-web:	http://AmericanMold.com/~ame/
Phone:	(612) 571-8642
Fax:	(612) 571-9393
Mailing address:	7230 Commerce Circle W.N.E., Fridley, MN 55432

American Physical Society (APS)

Description:	A professional organization for physicists. The organization has more than 41,000 physicists worldwide. Since its formation in 1899, it has been dedicated to the advancement and diffusion of the knowledge of physics.
Internet via-web:	http://www.aps.org/
Phone:	(301) 209-3224

American Public Works Association (APWA)

Description:	A professional organization. It is an association of public and private sector professionals who work in city, county, and state governments or for private companies. APWA focuses on transportation, mass transit, public buildings and parks, solid waste, power utilities, water treatment, and civil engineering/careers.
Internet via-web:	http://www.pubworks.org/apwa/main.html
E-mail:	apwa@bbs.pubworks.org;
Mailing address:	106 W. 11th Street, Suite 1800, Kansas City, MO 64105

American Turbine Pump Co.

Description:	A manufacturer of turbine pumps. Information includes product sheet, warranty, sample spec., etc.
Internet via-web:	http://204.49.131.2/atp/atphome.htm
Phone:	(806) 747-3248
Fax:	(806) 747-4329
Mailing address:	4229 Adrian Street, Lubbock Texas 79415

American Vacuum Society (AVS)

Description: A professional organization for the development of vacuum techniques. It is a nonprofit professional society dedicated to advancing the science and technology of films and coatings, microelectronics, nanostructures, surfaces and interfaces, plasmas, vacuum, and manufacturing processes.

Internet via-web: http://www.vacuum.org/

E-mail: angela@vacuum.org

Phone: (212) 248-0200

American Water Works Association (AWWA)

Description: AWWA is an international nonprofit scientific and educational society dedicated to the improvement of drinking water quality and supply. Founded in 1881, AWWA is the largest organization of water supply professionals in the world. Its more than 50,000 members represent the full spectrum of the drinking water community: treatment plant operators and managers, scientists, environmentalists, manufacturers, academicians, regulators, and others who hold genuine interest in water supply and public health. Membership includes more than 3,700 utilities that supply water to roughly 170 million people in North America.

Internet via-web: http://www.awwa.org/

E-mail: cberberi@awwa.org

Phone: (303) 794-7711

Mailing address: 6666 West Quincy Ave., Denver, CO 80235

Amoco Corporation

Description: A major oil company. Items include onstream magazine, explore our world, what we do, and master index.

Internet via-web: http://www.amoco.com/

Analytical Chemistry and Chemometrics Index

Description: An information hub on analytical chemistry and chemometrics areas. It links to analytical chemistry basics, analytical chemistry resources, chemometrics ar Aberystwyth, Neural networks, chemometrics ar Bristol, etc.

Internet via-web: http://www.chemie.fu-berlin.de/chemistry/index/anal/

Analytical Chemistry and Instrumentation

Description: A provider of information on analytical chemistry and instrumentation. The information includes data acquisition and electronics, data handling, diffraction methods, electrochemistry, gravimetric analysis, imaging techniques, mass spectrometry, materials and surface analysis, sensors, separations, spectroscopy, standards, thermal methods, and titration.

Internet via-web: http://www.chem.vt.edu/chem-ed/analytical/ac-methods.html

Mailing address: Department of Chemistry, Virginia Polytechnic Institute and State University, Blacksburg, VA 24061-0212

Analytical Chemistry Basics

Description: A basic analytical chemistry course offered by the Virginia Tech Chemistry Hypermedia project. Its contents include chemical equilibrium gravimetric analysis, titration, etc.

Internet via-web: http://www.chem.vt.edu/chem-ed/analytical/ac-basics.html

Mailing address: Department of Chemistry, Virginia Polytechnic Institute and State University, Blacksburg, VA 24061-0212

Analytical Chemistry Hypermedia

Description: Hypermedia tutorial materials that are used to supplement educational resources for students and practitioners of analytical chemistry.

Internet via-web: http://www.chem.vt.edu/chem-ed/analytical/ac-home.html

Mailing address: Department of Chemistry, Virginia Polytechnic Institute and State University, Blacksburg, VA 24061-0212

Analytical Service Laboratories (ASL)

Description: A consulting firm that provides full-service chemical testing, research, and consulting to environmental professionals since 1982.
Internet via-web: http://www.asl-labs.bc.ca/
E-mail: info@asl-labs.bc.ca

Analyticon Instruments Corporation

Description: An information resource hub. Provides information for manufacturer and supplier directory, global product source guide, scientific supply houses/catalog dealers and manufacturer representatives, trade shows, trade journal, publications and reports, organizations and scientific employment agents, and news groups.
Internet via-web: http://www.analyticon.com
E-mail: webmaster@analyticon.com
Phone: (201) 379-6711
Fax: (201) 379-6795
Mailing address: PO Box 92, 2 Edison Place, Springfield, NJ 07081

Applied Coatings and Linings, Inc.

Description: A supplier of coating and lining systems on metal parts manufactured or routed through Southern California.
Internet via-web: http://www.corrosion.com/applied/index.html
E-mail: applied@corrosion.com
Phone: (818) 280-6354
Fax: (818) 288-3310
Mailing address: 3224 N. Rosemead Blvd., El Monte, CA 91731

ARCO Chemical

Description: A manufacturer of chemicals including (1) ARCAL, which is used in paints, caulkings, adhesives and inks; (2) MPDiol, which is a fast-reacting liquid diol with unique properties for personal care, powder coatings, urethanes and adhesive products; (3) and NMP, which is the safer, recyclable, effective solvent.
Internet via-web: http://pubs.acs.org/pin/arco/arc.html

Argus Chemicals

Description: A manufacturer of various chemicals. The products include rare and fine chemicals, phosphorus organocompounds, and asymmetric synthesis.

Internet via-web: http://www.texnet.it/argus/argus.html

E-mail: argus@texnet.it

Phone: +39 574-938255

Fax: +39-574-950880

Mailing address: Via del Molino Nuovo, 2/A, I-50040 Mercatale di Vernio (PO)

ARSoftware's Online Internet Catalog

Description: A computer software company. It provides softwares on (1) thermodynamics and kinetics modeling, (2) molecular modeling and drawing, (3) compuDrug, (4) chemical engineering, (5) optical, (6) consulting, and (7) training.

Internet via-web: http://arsoftware.arclch.com/

E-mail: ars@ari.net

ASB Industries

Description: A plastics engineering company that specializes in industrial protective coatings.

Internet via-web: http://www.polysort.com/ASB

Phone: (216) 753-8458

Fax: (216) 753-7550

Mailing address: 1031 Lambert Street, Barberton, OH 44203

ASD

Description: A manufacturer of portable spectrophotometers, spectroradiometers and NIR analyzers.

Internet via-web: http://pluto.njcc.com/~bpapp/chempoin.html

ASEE Clearinghouse for Engineering Education

Description: ASEE is the American Society for Engineering Education. It is a nonprofit organization of individuals, institutions, and companies dedicated to improving all aspects of engineering education.

Internet via-web: http://www.asee.org/index.html

E-mail: webmaster@asee.org

Aslchem International Inc.

Description: A chemical importer/exporter of a Canadian firm.

Internet via-web: http://www.iceonline.com/home/aslchem

Mailing address: Vancouver, Canada

Aspen Technology

Description: A provider of software products including process modeling and process information management technology.

Internet via-web: http://www.aspentec.com/

E-mail: webmaster@aspentec.com

Associated Rubber Company

Description: A plastics engineering company that provides a full-service custom mixer and tread rubber manufacturer. The company employs over two-hundred and forty people at seven plants in Georgia, Alabama, Tennessee, and South Carolina.

Internet via-web: http://www.mindspring.com/~asscrc/

Phone: (770) 574-2321; (800) 277-8231

Mailing address: PO Box 245, 2130 U.S. Hwy. 78, Tallapoosa, GA 30176

ATI Cahn Company

Description: A plastics engineering company that provides a variety of services including instrumentation, micro and laboratory balances, etc.

Internet via-web: http://www.netopia.com:80/aticahn/

E-mail: larryg@atina.com

Phone: (608) 831-5515; (800) 423-6641

Fax: (608) 831-2093

Mailing address: 1001 Fourier Drive, Madison, WI 53717

ATI Orion

Description: A supplier of PerpHecT pH meters featuring digital LogR(tm) temperature compensation technology.

Internet via-web: http://pubs.acs.org/pin/orion/ori.html

Atlantis Plastics, Inc.

Description: A plastics engineering company that provides a variety of services for the plastics industry.

Internet via-web: http://www.cfonews.com/agh/

E-mail: webmaster@cfonews.com

Phone: (305) 858-2200

Augias Environmental Corp.

Description: A consulting firm that promotes and sells a surface cleaner degreaser, an environmentally friendly cleaner. It utilizes natural microbes that multiply rapidly and digest the hydrocarbons transforming them into carbon dioxide and water.

Internet via-web: http://www.pronett.com/augias/augias.htm

E-mail: Augias@aol.com

Phone: (703) 471-4952

Fax: (301) 229-1378

Mailing address: 13884 Park Center Road, Herndon, Virginia 22071

Ausimont USA, Inc.

Description:	A manufacturer of many chemical products. The company's products include PVDF, ECTFE, PTFE, MFA/PFA, elastomers, lubricants, fluids, and order Info.
Internet via-web:	http://Ausiusa.inter.net/ausiusa/
E-mail:	us008647@interramp.com
Phone:	(800) 221-0553; (609) 853-8119
Fax:	(609) 853-8711
Mailing address:	10 Leonards Lane, Thorofare, NJ 08086

Automotive Recycling Mailing List

Description:	This program is an international mailing list with the purpose of promoting the flow of information and knowledge. It is not for commercial purposes; advertisements are not allowed.
Internet via-web:	http://ie.uwindsor.ca/autorecy/welcome.html
E-mail:	spicer@uwindsor.ca

Baltzer Science Publishers

Description:	A publisher of many subjects. Information includes catalogue of journals, ordering information, author instructions, and publication schedule 1995.
Internet via-web:	http://www.nl.net/~baltzer
E-mail:	publish@baltzer.nl
Mailing address:	PO Box 8577, Red Bank, NJ 07701-8577

Basel Convention

Description: The Basel Convention on the Control of Transboundary Movements of Hazardous Wastes and Their Disposal was adopted in 1989. The treaty seeks to regulate the import and export of hazardous wastes to ensure that they are managed in an environmentally sound manner. The secretariat of the Basel Convention is administered by the United Nations Environment Programme

Internet via-web: http://www.unep.ch/sbc/about.html

Phone: (41 22) 979 9111

Fax: (41 22) 797 3454

Mailing address: Secretariat for the Basel Convention (SBC), UNEP, Geneva Executive Center, 15 chemin des Anemones, Building D, 1219 Chatelaine (Geneva), Switzerland

BASF Corporation

Description: A leading manufacturer of chemicals and related products from raw materials and auxiliaries through precursors and intermediates to finished goods. The BASF Group of companies have more than 100,000 employees worldwide that manufacture approximately 8,000 products from sites in 39 countries.

Internet via-web: http://www.basf.com

Mailing address: Headquartered in Ludwigshafen, Germany

Beckman Instruments, Inc.

Description: A manufacturer of a range of analytical instruments. Their products also include bioresearch and clinical diagnostics.

Internet via-web: http://www.beckman.com/

Phone: (800) 742-2345

Mailing address: 2500 Harbor Blvd., Box 3100, Fullerton, CA 92634

Beilstein Information Systems

Description: A German firm that specializes in organic chemistry. Information includes the company, press release, products, etc.

Internet via-web: http://www.beilstein.com/

E-mail: helpdesk@beilstein.com

Phone: (800) 275-6094; (303) 792-2652

Fax: (303) 792-2828

Mailing address: 15 Inverness Way East, Edgewood, CO 80112

Bifurcation and Nonlinear Instability Laboratory

Description: A research institute funded by NSF. Their research activities include pattern formation in Marangoni and Rayleigh type convection and teaching experiments in instability.

Internet via-web: http://gibbs.che.ufl.edu/bifurcation.shtml

E-mail: postmaster@gibbs.che.ufl.edu

Phone: (904) 492-0862

Mailing address: Department of Chemical Engineering, University of Florida, Gainesville, FL 32611

Bio Control Network

Description: A supplier of sustainable pest control alternatives for agriculture, horticulture, commercial, and residential use. BICONET offers a variety of preventative management resources including bio-intensive pest control, organic farm and garden products, educational materials, and global community feedback.

Internet via-web: http://www.usit.net/BICONET

Bio-Online

Description: A uniform resource locator (URL) directory. Includes a variety of server lists related to Biotechnology Industry Organizations (BIO).

Internet via-web: http://www.bio.com/companies/co-info.toc.html

E-mail: bio@bio.org

Phone: (202) 857-0244

Fax: (202) 857-0237

Mailing address: 1625 K Street, NW, Suite 1100, Washington, D.C. 20006-1604

BioSupplyNet

Description: A provider of an on-line directory for biomedical research products and services. Their services include automated search capabilities, new on the market, special offers, product user groups, automated inquiry functions, and bio-medical web servers.

Internet via-web: http://www.biosupplynet.com/bsn/

E-mail: doncorn@biosupplynet.com

Black & Veatch

Description: A global engineering and construction firm that specializes in energy, environment, process, and buildings. Clients include utilities, commerce, industry, and government agencies in more than 40 countries throughout the world.

Internet via-web: http://www.bv.com/

E-mail: info@bv.com

Phone: (913) 339-2000

Mailing address: 1500 Meadow Lake Pkway, Kansas, MO

Boulder Scientific Co.

Description:	A manufacturer of various chemicals. Specific chemicals of interest can be found by searching through the Internet homepage, WWW Chemicals, at the Internet address, "http://www3.ios.com:80/~ilyak."
Internet via-web:	http://www3.ios.com:80/~ilyak/ind019.html
Phone:	(303) 442-1199
Fax:	(303) 535-4584
Mailing address:	PO Box, 598 Third Street, Mead, CO 80542

BUBL Information Service

Description:	A uniform resource locator (URL) directory. Includes information in the areas of chemical engineering and research institutes. It currently contains sixty (66) servers from the United States and foreign countries.
Internet via-web:	http://www.bubl.bath.ac.uk/BUBL/Chemeng.html

Buckman Laboratories

Description:	A specialty chemical company serving the pulp and paper, water treatment, leather, coatings, agricultural, and wood treatment industries.
Internet via-web:	http://www.buckman.com/
Phone:	(901) 278-0330
Fax:	(901) 276-5343
Mailing address:	1256 North McLean Blvd., Memphis, TN 38108-0305

Burton Hamner's List of Internet Environmental Sources (Feb 17, 1995)

Description:	A uniform resource locator (URL) directory. Includes information related to green engineering and environmentally friendly activities.
Internet via-web:	http://ie.uwindsor.ca/ecdmlist/feb1995.2.html#feb179503

Calgon Corp.

Description: A manufacturer of specialty chemicals and a provider of services for water treatment. Their products include cooling water treatment, boiler treatment, wastewater treatment, municipal water treatment, paper process chemicals, surface treatment, cosmetic ingredients, and specialty biocides.

Internet via-web: http://www.calgon.com

E-mail: erdner1@calgon.com

Phone: (412) 777-8133 for new Calgon surround 1017 control; 800-955-0090 for softwater corrosion control product; (412) 777-8000 for ECLIPSE product

Fax: (412) 777-8927

Mailing address: PO Box 1346, Pittsburgh, PA 15230

CambridgeSoft, Corp.

Description: A computer software company that develops, markets, and supports desktop applications for chemists and engineers on, for example, (1) chemical structure drawing (ChemDraw), (2) molecular modeling/analysis (Chem3D), (3) chemical information management (ChemFinder).

Internet via-web: http://www.camsci.com/

E-mail: info@camsoft.com

Phone: (617) 491-2200

Fax: (617) 491-8208

Mailing address: 875 Massachusetts Ave., Cambridge, MA 02139

Carbolabs, Inc.

Description: A manufacturer of various chemicals. Specific chemicals of interest can be found by searching through the Internet homepage, WWW Chemicals, at the Internet address, "http://www3.ios.com:80/~ilyak."

Internet via-web: http://www3.ios.com:80/~ilyak/ind018.html

Phone: (203) 393-3029

Fax: (203) 393-2437

Mailing address: PO Box 3765 Amity Station, New Haven, CT 06525-0765

Carbon Dioxide Information Analysis Center (CDIAC)

Description: CDIAC provides information to help international researchers, policymakers, and educators evaluate complex environmental issues, including potential climate change associated with elevated levels of atmospheric carbon dioxide and other radiatively active trace gases. It also contains the World Data Center-A for Atmosphere Trace and Information System. CDIAC is a component of the U.S. Global Change Data and Information System (GCDIS).

Internet via-web: http://cdiac.esd.ornl.gov

E-mail: cdiac@ornl.gov

Phone: (423) 574-0390

Mailing address: Oak Ridge National Laboratory, PO Box 2008, Oak Ridge, TN 37831-6335

Catalytica, Inc.

Description: A manufacturer of many products. The company is currently focused on improving production of pharmaceutical intermediates and developing advanced combustion systems to reduce emissions generated by gas turbines. Their products include (1) xonon flameless combustion system, (2) gas monitors, (3) fine chemicals, and (4) consulting and R&D services.

Internet via-web: http://www.catalytica-inc.com/

E-mail: webmaster@catalytica-inc.com

Phone: (415) 960-3000

Fax: (415) 960-0127

Mailing address: 430 Ferguson Dr., Mountain View, CA 94043

Celgene Corp.

Description: A manufacturer of various chemicals. Specific chemicals of interest can be found by searching through the Internet homepage, WWW Chemicals, at the Internet address, "http://www3.ios.com:80/~ilyak."

Internet via-web: http://www3.ios.com:80/~ilyak/ind012.html

Phone: (908) 273-1001

Fax: (908) 271-4184

Mailing address: 7 Powder Horn Drive, Warren, NJ 07059

Center for Disease Control (CDC) and Prevention

Description: An agency of the Public Health Service within the Department of Health and Human Services. Their mission is to promote health and quality of life by preventing and controlling disease, injury, and disability.

Internet via-web: http://www.cdc.gov/

Internet via-gopher: gopher://gopher.cdc.gov

E-mail: netinfo@cdc1.cdc.gov

Mailing address: 1600 Clifton Road, NE, Atlanta, GA 30333

Center on Polymer Interfaces and Macromolecular Assemblies (CPIMA)

Description: CPIMA was established in September 1994 as one of eleven Materials Research Science and Engineering Centers (MRSECs) supported by the division of Materials Research of the National Science Foundation. CPIMA is a university/industrial partnership among research groups from Stanford University, IBM Almaden Research Laboratory, and the University of California Davis. CPIMA is devoted to the fundamental study of the interfacial science of organic thin films prepared from polymers and low-molecular weight amphiphiles. Potential applications lie in areas of linear and nonlinear optical devices, displays, electro-optical devices, sensors, permselective membranes, lubrication, and adhesion.

Internet via web: http://www.engr.ucdavis.edu/~chmsweb/faculty/stroeve/cpima/index.html

Ceramics and Industrial Minerals Home Page

Description: A uniform resource locator (URL) directory that hosts a variety of servers involved in ceramics, industrial minerals, and related raw materials and equipment.

Internet via-web: http://www.minerals.com/~ceramics/

E-mail: jmassari@ceramics.com

CFD Resources Online

Description:	A uniform resource locator (URL) directory that hosts a variety of servers in the area of computational fluid dynamics. Servers provided include turbulence, multigrid methods, and mesh generation.
Internet via-web:	http://www.tfd.chalmers.se/CFD_Online/
E-mail:	jola@tfd.chalmers.se

CH2M Hill

Description:	An engineering company providing water, environmental, transportation, and infrastructure services. The company has locations in over 70 cities throughout the world.
Internet via-web:	http://www.ch2m.com/
Phone:	(303) 771-0900

Challenge, Inc.

Description:	A chemical manufacturer specializing in biodegradable polymer barrier coatings for process paint stripping. Their products are used in decorative painting in the automotive, computer, electronics, and television markets. Because of their well-conducted quality control, the company has earned the ISO 9002 certification. ISO means the International Organization for Standardization. It is a worldwide federation of national standards bodies from some 100 countries, one from each country.
Internet via-web:	http://www.in.net/chem
E-mail:	mktg@challenge-inc.com
Phone:	(800) 548-7148
Fax:	(317) 876-1103
Mailing address:	7950 Georgetown Road, Suite 200, Indianapolis, IN 46268

ChemConnect

Description: A provider of a world directory for chemical buyers, manufacturers, and distributors. Their business covers chemicals products, manufacturers and distributors, custom manufacturers, trade names, chemicals wanted bulletin board, chemicals and services offered bulletin board.

Internet via-web: http://www.ChemConnect.com/

E-mail: webmaster@chemsource.com

Chemical Abstract Service (CAS)

Description: A division of the American Chemical Society. CAS provides a broad range of scientific information products, from focused topical publications to comprehensive online sources. It is the publisher of the printed *Chemical Abstracts*.

Internet via-web: http://info.cas.org/about.html

E-mail: help@cas.org

Phone: (614) 447-3600

Fax: (614) 447-3713

Mailing address: 2540 Olentangy River Road, Columbus, OH 43210

Chemical Concepts

Description: A German firm. It is a manufacturer-independent supplier of spectroscopy software ranging from companywide installations to individual, desktop programs.

Internet via-web: http://www.vchgroup.de/cc/

E-mail: marketing@cc.vchgroup.de

Phone: 49 (6201) 606 433

Fax: 49 (6201) 606 430

Mailing address: Boschstrasse 12, D-69469 Weinheim, Germany

Chemical Education Resources (CER)

Description:	A publisher of individual laboratory experiments. No minimum quantity or number of titles is required for class-size orders.
Internet via-web:	http://www.cerlabs.com/chemlabs
Phone:	(717) 838-3165
Fax:	(717) 838-6275
Mailing address:	PO Box 357, Palmyra, PA 17078

Chemical Engineering Sites All Over the World

Description:	An academic uniform resource locator (URL) directory that hosts a variety of servers in the area of chemistry and chemical engineering academic institutions all over the world. The information is maintained by Stefan Gerl, University of Karlsruhe.
Internet via-web:	http://www.ciw.uni-karlsruhe.de/chem-eng.html
Mailing address:	Chemical Engineering Department, University of Karlsruhe, Kaiserstrasse 12, 76128 Karlsruhe, Germany

Chemical Engineering: Information Indexes

Description:	A world-wide web virtual library containing lists of information resources relevant to chemical and process engineering.
Internet via-web:	http://www.che.ufl.edu/www-che/topics/indexes.html
E-mail:	www-che@www.che.ufl.edu
Phone:	(904) 492-0862
Mailing address:	Department of Chemical Engineering, University of Florida, Gainesville, FL 32611

Chemical Engineering: Professional Organization

Description:	A world-wide web virtual library containing lists of information resources relevant to chemical and process engineering.
Internet via-web:	http://www.che.ufl.edu/www-che/topics/research.html
E-mail:	www-che@www.che.ufl.edu
Phone:	(904) 492-0862
Mailing address:	Department of Chemical Engineering, University of Florida, Gainesville, FL 32611

Chemical Engineering: Research Organization and Laboratories

Description:	A world-wide web virtual library containing lists of information resources relevant to chemical and process engineering.
Internet via-web:	http://www.che.ufl.edu/www-che/topics/research.html
E-mail:	www-che@www.che.ufl.edu
Phone:	(904) 492-0862
Fax:	not found
Mailing address:	Department of Chemical Engineering, University of Florida, Gainesville, FL 32611

Chemical Marketing Online (CHEMON)

Description:	A provider of an online marketplace for various chemicals. Their business includes chemicals for sale, chemical wanted, equipment, services, companies, etc.
Internet via-web:	http://www.chemon.com/
E-mail:	chemon@mkts.com
Fax:	(914) 763-4263
Mailing address:	PO Box 246, Pound Ridge, NY 10576

Chemical Physics Preprint Database

Description:	A uniform resource locator (URL) directory that hosts a variety of servers for the international theoretical chemistry community. It is maintained by the Brown University and the Theoretical Chemistry and Molecular Physics Group at the Los Alamos National Laboratory.
Internet via-web:	http://www.chem.brown.edu/chem-ph.html

Chemical Process Modeling and Flowsheet Synthesis

Description:	About an individual person who is interested in conducting seminars on the subject of chemical process modeling and flowsheet synthesis.
Internet via-web:	http://www.preferred.com/~lpartin/index.html
E-mail:	lpartin@preferred.com

Chemical Week Magazine

Description:	The magazine publishes 48 times each year. A uniform resource locator (URL) directory that hosts a variety of servers for chemical makers and processors. The servers provided include worldwide chemical industry, advertising opportunities to reach a global audience, chemical week industry conferences, buyer's guide, and hot links.
Internet via-web:	http://www.chemweek.com/
Phone:	(212) 621-4900
Mailing address:	888 Seventh Ave., Twenty-Sixth Floor, NY, NY 10106

ChemInnovation Software

Description:	A provider of computer software for analytical chemistry. Their business covers chemistry 4-D draw, namexpert, new windows 95 vision, and IUPAC nomenclature rules.
Internet via-web:	http://www.cheminnovation.com/
E-mail:	cis@ChemInnovation.com
Phone:	(619) 566-2846
Fax:	(619) 566-4138
Mailing address:	8190-E Mira Mesa Blvd, #108., San Diego, CA 92126

Chemistry

Description:	A virtual library. It provides (1) WWW Chemistry Sites at Academic Institutions, (2) WWW Chemistry Sites at Nonprofit Organizations, (3) WWW Chemistry Sites at Commercial Organizations, (4) Other Lists of Chemistry Resources and Related WWW Virtual Libraries, (5) Some Chemistry Gopher Servers, (6) Some Chemistry FTP Servers, (7) Chemistry and Biochemistry USENET News Groups.
Internet via-web:	http://www.chem.ucla.edu/chempointers.html
E-mail:	mik@chem.ucla.edu

Chemistry Index / Chemie-Index (FU Berlin)

Description: A uniform resource locator (URL) directory that hosts a variety of servers for chemists. The servers provided include general chemistry, biochemistry, organic chemistry; chemicals, safety, etc.

Internet via-web: http://www.chemie.fu-berlin.de/chemistry/index.html

Chemistry on the Internet

Description: Selected as "the best of the Web 1995" in the 1995 ACS Symposium in Chicago, Illinois. It is a uniform resource locator (URL) directory that contains a variety of servers in the area of chemistry and chemistry related field.

Internet via-web: http://www.ch.ic.ac.uk/infobahn/boc.html

Chemistry Sites at Commercial Organizations

Description: A world-wide web virtual library containing lists of chemistry resources relevant to commercial organizations.

Internet via-web: http://www.chem.ucla.edu/chempointers.html#www_commercial

E-mail: mik@chem.ucla.edu

Chemistry Sites at Nonprofit Organizations

Description: A world-wide web virtual library containing lists of chemistry resources relevant to nonprofit organizations.

Internet via-web: http://www.chem.ucla.edu/chempointers.html#www_nonprofit

E-mail: mik@chem.ucla.edu

Chemistry Sites at Other Information Resources

Description: A world-wide web virtual library containing lists of chemistry resources relevant to other information resources.

Internet via-web: http://www.chem.ucla.edu/chempointers.html#www_vl

E-mail: mik@chem.ucla.edu

ChemKey Database

Description: An Organic Synthetic Method Computer Database (45,000 References Available). The features of the database include easy operation, boolean logic, etc.

Internet via-web: http://euch6f.chem.emory.edu/

E-mail: chemap@dooley.cc.emory.edu

Fax: (404) 727-6629

Mailing address: Heterodata, Inc., 1055 Rosewood Drive, Atlanta, GA 30306

ChemSearch (Chemical Recycling)

Description: A provider of an online marketplace for various chemicals. Their business covers quick link--chemicals, quick link--your needs or usage, quick link--your surplus or excess, quick link--recycling forum.

Internet via-web: http://www.sonic.net/chemsearch

E-mail: paulp@sonic.net

Phone: (707) 823-4331

Fax: (707) 823-1477

Mailing address: 321 South Main Street, No. 524, Sebastopol, CA 95472

ChemSOLVE

Description: An environmental analytical laboratory, specializing in analytical services, and providing reference software for the environmental professional.

Internet via-web: http://www.eden.com/~chemsolv/

E-mail: chemsolve@aol.com

Phone: (512) 280-7680

Fax: (512) 280-7651

Mailing address: 11629 Manchaca Road, Austin, TX

Chemsyn Science Lab.

Description: A manufacturer of various chemicals. Specific chemicals of interest can be found by searching through the Internet homepage, WWW Chemicals, at the Internet address, "http://www3.ios.com:80/~ilyak."

Internet via-web: http://www3.ios.com:80/~ilyak/ind017.html

Phone: (800) 233-6643

Fax: (913) 888-3582

Mailing address: 13605 W. 96th Terrace, Lenexa, KS 66215-1297

Chemtec Publishing

Description: A publisher of chemically related subjects. Their business covers journals, books, and software.

Internet via-web: http://www.io.org/~chemtec/

E-mail: chemtec@io.org

Phone: (416) 265-2603

Mailing address: 38 Earswick Dr., Toronto-Scarborough, Ontario M1E 1C6 Canada

Chevron

Description: A major oil company. The Chevron ranks as the fifth-largest oil company in the world (based on revenues), the largest U.S. marketer of petroleum products, one of the largest marketers of liquefied petroleum gas worldwide, and the third-largest U.S. producer of natural gas.

Internet via-web: http://www.chevron.com/

Chromophore, Inc

Description:	A manufacturer of polymers and small molecules for use in nonlinear optics, biological staining, fluorescent tagging, and thermochromic dyes.
Internet via-web:	http://www.chromophore.com/
Phone:	(205) 533-6610
Fax:	(205) 533-4805
Mailing address:	2307 Spring Branch Rd., Huntsville, AL 35801

Chrysler Corp. Recycling

Description:	Addresses replacing CFCs and ozone-depleting chemicals in Chrysler's products and plants. Chrysler is also minimizing waste through recycling and conservation.
Internet via-web:	http://www.chryslercorp.com/environment/recycling.html

Chugoku Kogyo Co.,Ltd

Description:	A provider of inorganic chemicals, nonferrous metals and precious metals, organic chemicals. Their business includes recovery and sales of spent catalysts and scraps of rare metals.
Internet via-web:	http://chemical-metal.co.jp/cgk/

Ciba-Geigy AG Basel

Description:	A biological and chemical company, based in Switzerland. Their business covers health care, agriculture, agriculture and industry with innovative value-adding products and services.
Internet via-web:	http://147.167.128.11/
E-mail:	http://www.ciba.com

Citylink

Description:	A uniform resource locator (URL) directory that hosts a variety of servers for all U.S. states and many U.S. cities.
Internet via-web:	http://www.NeoSoft.com/citylink/
E-mail:	citylink@neosoft.com
Phone:	(504) 898-2158
Fax:	(504) 892-8535

Civil Engineering

Description:	A uniform resource locator (URL) directory that hosts a variety of servers in the area of civil engineering. The servers provided include universities, organizations, government agencies, and commercial homepages from around the world.
Internet via-web:	http://www.ce.gatech.edu/WWW-CE/home.html
E-mail:	nelson.baker@ce.gotech.edu
Mailing address:	School of Civil Engineering and Environmental Engineering, Georgia Tech., Atlanta, GA

CLI International, Inc.

Description:	A provider of contract services for materials evaluation, corrosion testing, consulting and failure analysis to industry. Their business covers corrosioneering newsletter, software, training, engineering services, corrosion research, and inhibitor evaluation.
Internet via-web:	http://www.clihouston.com/
Phone:	(713) 444-2282
Fax:	(713) 444-0246
Mailing address:	Bammel-N Houston, Suite 300, Houston, TX 77014

Clorox Company

Description:	A manufacturer of many chemical products including laundry additives, home cleaning products, cat litters, insecticides, charcoal briquets, food products, and water filter systems.
Internet via-web:	http://www.clorox.com/
E-mail:	Info@clorox.com
Mailing address:	PO Box 24305, Oakland, CA 94623-1305

Coatings Industry

Description:	Formed to provide common marketing, sales, environmental and technology transfer among noncompeting companies in the paint coating, adhesive, and basic resin manufacturing industries.
Internet via-web:	http://www.coatings.org/cia/
E-mail:	inforcia@coatings.org

Communications for a Sustainable Future

Description:	A uniform resource locator (URL) directory that hosts a variety of servers related to mailing lists of green engineering.
Internet via-web:	dhttp://csf.colorado.edu/

Communicopia Environmental Research and Communications

Description:	A consulting firm that provides Internet consultation and media services for the environmental and natural resource sectors of the economy.
Internet via-web:	http://www.communicopia.bc.ca

Computer-Aided Process Design Consortium (CAPD)

Description:	An industrial body within the Department of Chemical Engineering and the Engineering Design Research Center that deals with the development of methodologies and computer tools for the process industries. The CAPD consortium currently has more than 20 members from chemical and petroleum companies, and from a number of hardware and software companies.
Internet via-web:	http://www.cheme.cmu.edu/research/capd/
Phone:	(412) 268-3372
Mailing address:	The Engineering Design Research Center, Carnegie Mellon University, 5000 Forbes Avenue, Pittsburgh, PA 15213

Consortium on Green Design and Manufacturing

Description: An interdisciplinary research initiative at the University of California, Berkeley and an industry/government/university partnership. The goal is to develop linkages between manufacturing and design and their environmental effects and to integrate engineering information, management practices, and government policy-making.

Internet via-web: http://euler.berkeley.edu/green/cgdm.html

CS Distribuidora, S.A. de C.V.

Description: A manufacturer of products destined to the chemical-pharmaceutical, plastics, and electronics industries. Their products include dicadox, floxal, fostiacil, fostiacil-c, furadim, furobac, porzonidazol, pulmocin, pulmoxisul, and trimexol.

Internet via-web: http://www.spin.com.mx/grupocs/gcs-csdb.html

CTD, Inc

Description: A developer of cyclodextrin applications. Information includes products and services, cyclonet, order form, etc.

Internet via-web: http://www.cyclodex.com

Fax: (904) 375-8287

Mailing address: Cyclodextrin Technologies Development, Inc., 3713 SW 42nd Avenue, Suite 3, Gainesville, FL 32608-2531

Custom Plastic Extrusions

Description: A plastics engineering company that provides service on plastic extrusions.

Internet via-web: http://www.polysort.com/companies/c/cpe/cpe.html

Phone: (216) 297-1426

Fax: (216) 297-9423

Mailing address: 347 Day Street, PO Box 409, Ravenna, Ohio 44266

Dalton Chemical Laboratories, Inc.

Description:	A chemical company specializing in DNA/RNA synthesis products. Their business covers DNA/RNA synthesis products, custom oligonucleotide synthesis, custom services and synthesis.
Internet via-web:	http://www.dalton.com/dalton
E-mail:	chemist@dalton.com
Phone:	(416) 736-5394; (800) 567-5060
Fax:	(416) 736-5846
Mailing address:	Farquharson Building, York University Campus, 4700 Keele Street, North York, Ontario M3J 1P3, Canada

Daylight Chemical Information Systems, Inc.

Description:	A provider of chemical information software. Their business covers Daylight ToolKits, application software, and database.
Internet via-web:	http://www.daylight.com/
E-mail:	info@daylight.com
Phone:	(714) 476-0451
Fax:	(714) 476-0654
Mailing address:	18500 Von Karman, STE #450, Irvine, CA 92715

Deepwater Iodides, Inc.

Description:	A manufacturer of various chemicals. Specific chemicals of interest can be found by searching through the Internet homepage, WWW Chemicals, at the Internet address, "http://www3.ios.com:80/~ilyak."
Internet via-web:	http://www3.ios.com:80/~ilyak/ind021.html
Phone:	(800) 854-4064
Fax:	(405) 256-0575
Mailing address:	1210 Airpark Road, Woodward, OK 73801

Department of Commerce (DOC)

Description: DOC was established on February 14, 1903 to promote American businesses and trade. Their broad range of responsibilities include expanding U.S. exports, developing innovative technologies, gathering and disseminating statistical data, measuring economic growth, granting patents, promoting minority entrepreneurship, predicting the weather, and monitoring stewardship. As diverse as Commerce's services are, there is a mandate that unifies them to work with the business community to foster economic growth and the creation of new American jobs.

Internet via-web: http://www.doc.gov

E-mail: stat-usa@doc.gov

Mailing address: Herbert C. Hoover Building, Washington, DC

Department of Energy (DOE)

Description: In partnership with customers, DOE is entrusted to contribute to the welfare of the Nation by providing the technical information and scientific and educational foundation for technology, policy, and institutional leadership necessary to achieve efficiency in energy use, diversity in energy sources, a more productive and competitive economy, improved environmental quality, and a secure national defense.

Internet via-web: http://www.doe.gov

E-mail: webmaster@apollo.osti.gov

Department of Energy, Office of Industrial Technologies (OIT)

Description: DOE has established the OIT Chemicals Industry Team, which is a part of the Industries of the Future strategy, to partner with the U.S. chemical industry to maximize economic, energy, and environmental benefits through research and development of innovative technologies. One of Team's activities is to develop environmentally friendly products such as "green" solvents.

Internet via-web: http://www.nrel.gov/oit/Industries-of-the-Future/chemical.html

E-mail: webmaster.oit@hq.doe.gov

Department of Health and Human Services (DHHS)

Description: DHHS is the United States government's principal agency for protecting the health of all Americans and providing essential human services, especially for those who are least able to help themselves. It includes:

- Public Health Service (PHS)
  - Agency for Toxic Substances and Disease Registry (ATSDR)
  - Centers for Disease Control and Prevention (CDC)
- Food and Drug Administration (FDA)
- National Institutes of Health (NIH)
  - National Institute of Environmental Health Sciences (NIEHS)
  - National Library of Medicine

Internet via-web: http://www.os.dhhs.gov/

E-mail: cgardett@os.dhhs.gov

Mailing address: Hubert H. Humphrey, 200 Independence Avenue, SW, Washington, DC 20201

Department of Labor (DOL)

Description: DOL is charged with preparing the American workforce for new and better jobs, and ensuring the adequacy of America's workplaces. It is responsible for the administration and enforcement of over 180 federal statutes. These legislative mandates and the regulations produced to implement them cover a wide variety of workplace activities for nearly 10 million employers and well over 100 million workers including protecting workers' wages, health and safety, employment and pension rights; promoting equal employment opportunity; administering job training, unemployment insurance and workers' compensation programs; strengthening free collective bargaining; and collecting, analyzing, and publishing labor and economic statistics.

Internet via-web: http://www.dol.gov

E-mail: webmaster@dol.gov

Design for the Environment at the US Department of Energy

Description: This program was supported by the Department of Energy, EM-334. It's aim is to develop an integrated set of tools to help engineers, designers, and planners incorporate pollution prevention strategies into the design stage of new products, processes, and facilities. P2 by Design is managed through the Pacific Northwest National Laboratory, which is operated by Battelle Memorial Institute for the U.S. Department of Energy. This project is a collaborative effort among Hanford contractors.

Internet via-web: http://w3.pnl.gov:2080/DFE/home.html

E-mail: sj_widener@pnl.gov

Phone: (509) 375-3703

Mailing address: Pacific Northwest National Laboratory, Battelle Blvd, MSIN K2-40, Richland, WA 99352

Diaz Chemical Corp.

Description: A manufacturer of various chemicals. Specific chemicals of interest can be found by searching through the Internet homepage, WWW Chemicals, at the Internet address, "http://www3.ios.com:80/~ilyak."

Internet via-web: http://www3.ios.com:80/~ilyak/ind024.html

Phone: (716) 638-6321

Fax: (716) 638-8356

Mailing address: PO Box 194, 40 Jackson Street, Holley, NY 14470

Dielectric Polymers Inc.

Description: A subsidiary of Park Electrochemical Corporation and a manufacturer of many chemical products including coating pressure sensitive rubber, acrylic, and silicone adhesives on release liners as free transfer films; double-adhesive coated paper, tissue, and film with a release liner; and in self-wound tape form coated on polyester, and polyamide film.

Internet via-web: http://www.dipoly.com/

E-mail: joe@dipoly.com

Phone: (413) 532-3288

Fax: (413) 533-9316

Mailing address: 218 Race Street, Holyoke, MA 01040

DMP Corporation

Description: A manufacturer of industrial wastewater treatment systems offering automated continuous-flow and batch treatment devices that handle wastewaters from a variety of industrial processes.

Internet via-web: http://web.sunbelt.net/dmp/dmp.htm

Phone: (803) 548-0853; (800) 845-3681

Fax: (803) 548-3590

Mailing address: PO Box 1088, Fort Mill, SC 29716

Dojindo Laboratories

Description: A manufacturer of analytical and biochemical reagents. The company has a collection of WWW homepages useful for chemical and biochemical researchers. Information includes companies, organizations, online journals, online services, mass communications, searchable homepages.

Internet via-web: http://www.dojindo.co.jp/

E-mail: info@dojindo.co.jp

Phone: 81-96-286-1515

Fax: +81-96-286-1525

Mailing address: Tabaru 2025-5, Mashiki-machi, Kamimashiki-gun Kumamoto 861-22, Japan

Dow Specialty Chemicals

Description: A division of Dow Chemical Company. Products--such as aqueous acrylimide monomer, diphenyl oxide, etc.--of the specialty chemicals are provided in the Internet homepage.

Internet via-web: http://www.dow.com:80/specialty/index.html

E-mail: dowcig@cris.com

Phone: (800) 447-4369

Fax: (517) 832-1465

DuPont

Description: A science and technology-based global company. The company provides either ingredients or end-products for markets such as aerospace, agriculture, apparel, chemicals, construction, electronic systems, energy, environment, food, healthcare, home products, leisure, packaging, publishing, pulp and paper, safety and transportation. Their major product areas include chemicals, fibers, films, finishes, petroleum, plastics, healthcare products, biotechnology, and composite materials.

Internet via-web: http://www.dupont.com

E-mail: info@dupont.com

Mailing address: E.I. du Pont de Nemous and Company, 1007 Market Street, Wilmington, Delaware 19898

E & D Plastics

Description: A plastics engineering company that provides fabrication services to the plastics industry.

Internet via-web: http://edp.ncc.com/edp/

E-mail: edp@onramp.net

Phone: (214) 742-6032

Fax: (214) 742-6087

Mailing address: Dallas, Texas

East Bay Municipal Utility District (EBMUD)

Description: The East Bay Municipal Utility District, Oakland, California, USA. EBMUD is a publicly owned water district formed in 1923 under the Municipal Utility District Act of 1921.

Internet via-web: http://www.ebmud.com/index.html

Phone: (510) 835-3000

Mailing address: 375 Eleventh Street, Oakland, CA 94607-4240

Eastern Minerals and Chemicals

Description: A consulting firm that provides brokerage, agency, and consulting services for a wide range of nonmetallic minerals.

Internet via-web: http://www.ceramics.com/~ceramics/emc/

Phone: (717) 295-7292

Fax: (717) 295-721

Mailing address: 147 North Shippen Street, Lancaster, PA 17604

Eastman Chemical Company

Description: A chemical manufacturing company that produces a wide range of chemicals, fibers, and plastics. Eastman polyethylene terephthalate (PET) plastic is used in containers for beverages, foods, and toiletries. Several brands of tools, toys, glasses, and toothbrushes feature products made from other Eastman plastics. Eastman products are in fabrics, floor covering, paints, pharmaceuticals, foods, cigarette filters, and chemical intermediates.

Internet via-web: http://www.eastman.com/

Phone: (423) 229-200

Fax: (423) 224-0413

Mailing address: PO Box 1974, Kingsport, TN 37662

ECDM Group, Michigan Technological University

Description: A research institute. The institute provides a list of servers that are related to green engineering. The servers provided include life cycle assessment.

Internet via-web: http://www.me.mtu.edu/research/envmfg/

Phone: (906) 487-3396

Mailing address: Michigan Technological University

Eco-Glass Group

Description: A manufacturing company that specializes in the secondary glass, fiberglass, and mineral by-products.

Internet via-web: http://www.recycle.net/recycle/Trade/rs000749.html

Ecocycle newsletter

Description: A newsletter publication company. Their goal is to promote life cycle tools, management and product policy. Maintained by Environment Canada.

Internet via-web: http://www.doe.ca/ecocycle/

Phone: (819) 997-3060

Fax: (819) 953-6881

Mailing address: Environment Canada, Hazardous Waste Branch, Ottawa, Ontario, Canada, K1A 0H3

EINET Galaxy

Description: An information search engine. Subjects that can be searched in this engine cover most human activities including arts, business, computers, education, engineering, entertainment, government, health, medicine, news, recreation, science, social science, society, and culture.

Internet via-web: http://galaxy.einet.net/

Electrochemical Society, Inc

Description: A professional organization concerned with a broad range of phenomena related to electrochemical and solid state science and technology.

Internet via-web: http://www.electrochem.org/ecs/

E-mail: ecs@electrochem.org

Phone: (609) 737-1902

Fax: (609) 737-2743

Mailing address: 10 South Main Street, Pennington, NJ 08534

Electronic Selected Current Aerospace Notices (E-SCAN)

Description: Listings include chemistry and materials (general), composite materials, inorganic and physical chemistry, metallic materials, nonmetallic materials, propellants and fuels, and materials processing.

Internet via-web: http://www.sti.nasa.gov/scan.html

E-mail: smullen@sti.nasa.gov

Phone: (301) 621-0320

Eli Lilly and Company

Description: A global research-based pharmaceutical corporation headquartered in Indianapolis, Indiana. Lilly pursues health care solutions by combining pharmaceutical innovation, the best of existing pharmaceutical technology, disease-prevention and disease management expertise, and information technologies.

Internet via-web: http://www.lilly.com/

E-mail: webmaster@lilly.com

Phone: (317) 276-2000

Mailing address: Indianapolis, IN 462855

Elsevier Science B.V.

Description: A publisher of scientific information. Their business includes electronic journals, earth, electrical and electronic engineering, mathematics and computer science, etc. It publishes more than 1100 English-language journals.

Internet via-web: http://www.elsevier.nl/

E-mail: nlinfo-f@elsevier.nl

Mailing address: PO Box 211, 1000 Ave, Amsterdam, The Netherlands

EMAX Solution Partners

Description: An information technology services company that provides enterprise systems solutions for chemical information and compliance management to major corporations.

Internet via-web: http://www.emax.com/

Energy & Environmental Research Center

Description: A coal research institute. Their programs include the areas of advanced power systems, waste disposal, waste reuse, air emissions control, biomass fuels, energy policy, contaminant cleanup, and mine land reclamation. Maintained by the University of North Dakota.

Internet via-web: http://www.eerc.und.nodak.edu/

E-mail: pmiller@eerc.und.nodak.edu

Phone: (701) 777-5000

Fax: (701) 777-5181

Mailing address: University of North Dakota, Box 9018, Grand Forks, ND 58202

Energy and Environmentally Conscious Manufacturing

Description: The Oak Ridge Centers for Manufacturing Technology (ORCMT) have identified that manufacturing processes with low energy and environmental costs are a key to enhanced competitiveness.

Internet via-web: http://www.ornl.gov/orcmt/energy/home.html

E-mail: 4usa@ornl.gov

Phone: (800) 356-4usa

Energy Federation, Inc. (EFI)

Description: A consulting firm that is promoting environmentally benign and sustainable use of energy and water resources. EFI helps to gain access to appropriate technologies and information for using energy and water efficiently.

Internet via-web: http://www.tiac.net/users/efi/

E-mail: info@efi.org

Phone: (508) 653-4299

Fax: (508) 655-3811

Mailing address: 14 Tech Circle, Natick, MA 01760

Engineered Rubber Products

Description: A plastics engineering company that specializes in the molding of quality natural and synthetic rubber parts.

Internet via-web: http://www.polysort.com/ERP

Phone: (216) 867-7700

Fax: (216) 867-3221

Mailing address: 1745 Copley Rd., Akron, OH 44320

Engineering Foundation

Description: A not-for-profit corporation. The Engineering Foundation was established in 1914 in response to Cleveland manufacturer Ambrose Swasey's offer to provide funds for research in science and engineering and the advancement of engineering in general. It awards Grants for Exploratory Research and Grants for the Advancement of Engineering, cosponsors the Engineering Journalism Award with the American Association of Engineering Societies, and conducts an extensive program of interdisciplinary conferences.

Internet via-web: http://www.engfnd.org/engfnd/

E-mail: engfnd@aol.com

Phone: (212) 705-7836

Fax: (212) 705-7441

Mailing address: 345 East 47th Street, Suite 303, New York, NY 10017

EnviroLink Network

Description: An environmental information clearinghouse. The company promotes a sustainable society through the use of new technologies.

Internet via-web: http://envirolink.org/

Enviromine

Description: A consulting firm that provides professionals and the interested public with information on environmental issues related to mining activities.

Internet via-web: http://www.info-mine.com/technomine/enviromine/env_main.html

E-mail: infodata@info-mine.com

Phone: (604) 683-2037

Fax: (604) 681-4166

Environment and Ecology

Description: A uniform resource locator (URL) directory that hosts a variety of gopher addresses. Their gopher menu includes: about this directory, agenda 21 hypertext version, army green, etc.

Internet via-gopher: gopher://riceinfo.rice.edu/11/Subject/Environment

Environment Canada

Description: A uniform resource locator (URL) directory that hosts a variety of servers related to mailing lists of green engineering. Maintained by the Environment Canada.

Internet via-web: http://www.doe.ca/

Phone: (819) 997-3060

Fax: (819)953-6881

Mailing address: Environment Canada, Ottawa, Ontario, Canada, K1A OH3

Environment One Corporation (E/ONE)

Description: A manufacturer and provider of (1) products and services for the disposal of residential sanitary waste and (2) detection systems for the protection of equipment and information. Their manufacturing, engineering, administration, and marketing operations are managed from Schenectady. Environment One also maintains regional sales offices in key cities in the U.S.A. and Japan.

Internet via-web: http://www.eone.com/eone/

E-mail: eonecom@aol.com

Phone: (518) 346-6161

Mailing address: Schenectady, New York

Environmental Concerns, Inc. (ECI)

Description: A supplier of environmental and water treatment products. The company markets water treatment systems and other environmental remediation technologies, organizes joint ventures, and assists in development of emerging environmentally oriented companies.

Internet via-web: http://www.datacor.com/~eci

E-mail: reesclark@datacor.com

Phone: (206) 391-1951

Fax: (206) 391-2009

Mailing address: 1065 Twelfth Ave., NW, Suite E1, Issaquah, WA 98027

Environmental Engineering

Description: A uniform resource locator (URL) directory that hosts a variety of servers in the area of environmental engineering. The servers provided include (1) universities with environmental engineering programs, (2) environmental engineering-related information, (3) environmental information (nonengineering), and (4) an invitation to submit items.

Internet via-web: http://www.nmt.edu/~jjenks/engineering.html

Environmental Industry Web Site

Description: A uniform resource locator (URL) directory that hosts a variety of servers which provide environmental services and products. Also provides information on ISO 14000 guides.

Keywords include environmental industry, environmental business, environmental services, environmental products, environmental engineering, environmental management, environment, environmental, industry, business, businesses, companies, company, services, products, engineering, consulting, cleanup, remediation, recycling, toxic, hazardous, waste, air, water, soil, pollution, pollution prevention, ISO14000, business opportunities, opportunities, green, organizations, associations, publications, magazines, directories, journals.

Internet via-web: http://www.enviroindustry.com/

E-mail: homepage@enviroindustry.com

Environmental Library

Description: A uniform resource locator (URL) directory that hosts a variety of servers in environmental areas.

Internet via-web: http://envirolink.org/envirowebs.html

Environmental News Network

Description: A publication company that provides a variety of environmental news. Information includes news, features, calendar, marketplace, library, etc.

Internet via-web: http://www.enn.com

Environmental Professional's Guide to the Net (EPGN)

Description: This guide is for environmental professionals worldwide who are interested in locating technical homepages on the Internet. More than just a hotlist of related links, the EPGN provides brief descriptions of each homepage to assist the user in evaluating a homepage prior to visiting it. This can save search time, particularly for those with slower modems.

Internet via-web: http://www.geopac.com/

E-mail: garnold@geopac.com

Environmental Protection Agency (EPA)

Description: EPA was established as an independent agency in the Executive Branch in December 1970 and is responsible for executing the federal laws protecting the environment. The Agency now administers nine comprehensive environmental protection laws, such as the Clean Air Act, the Clean Water Act, the Toxic Substances Control Act, the Resource Conservation and Recovery Act, and the Comprehensive Environmental Compensation and Liability Act (or "Superfund").

The EPA was created to permit coordinated and effective governmental action on the behalf of the environment. The EPA endeavors to systematically reduce and control pollution through the appropriate integration of a variety of research, monitoring, standard setting, and enforcement activities. The EPA also coordinates and supports research and antipollution activities by state and local governments, private and public groups, individuals, and educational institutions. In total, the EPA is designed to serve as the public's advocate for a livable environment.

The Agency is directed by an Administrator and a Deputy Administrator. The Agency's executive staff includes nine Assistant Administrators who manage specific environmental programs and direct other Agency functions, as well as Associate Administrators and the Agency's General Counsel and its Inspector General.

The ten Regional Offices across the country cooperate closely with state and local governments to make sure that federal environmental laws are properly implemented. They are also responsible for accomplishing within the Regions the national program objective established by the Agency.

Internet via-web: http://www.epa.gov/

Mailing address: 401 M Street, SW, Washington, DC 20460

EPA Acid Rain Hotline

Description: The Acid Rain Hotline records questions and disseminates EPA documents related to the Acid Rain Program. The Hotline assists callers who have specific technical questions by forwarding these inquiries to experienced EPA Acid Rain Division personnel, who review them and respond to the caller, usually within 24 hours. Utilities may find the Hotline especially useful for obtaining information that may help them comply with the acid rain regulations.

EPA's Acid Rain Hotline records requests for materials and refers technical questions to the Division.

Internet via-web: http://www.epa.gov/access/chapter3/s1-1.html

E-mail: internet_support@unixmail.rtpnc.gov

Phone: (617) 674-7377

Fax: (617) 674-2851

Mailing address: Acid Rain Hotline, c/o Eastern Research Group, 110 Hartwell Avenue, Lexington, MA 02173

EPA Air Risk Information Support Center Hotline

Description: The Air Risk Information Support Center Hotline has been developed to assist state and local air pollution control agencies and EPA Regional offices with technical matters pertaining to health, exposure, and risk assessment of air pollutants.

Internet via-web: http://www.epa.gov/access/chapter3/s1-2.html

E-mail: internet_support@unixmail.rtpnc.gov

Phone: (919) 541-0888

Fax: (919) 541-4028 or 2045

Mailing address: AIR RISC, U.S. Environmental Protection Agency, Office of Air Quality Planning and Standards,MD-13, or Environmental Criteria Assessment Office, MD-52, Research Triangle Park, NC 27711

EPA Alternative Treatment Technology Information Center

Description: ATTIC is a comprehensive information retrieval system containing data on alternative treatment technologies for hazardous waste. ATTIC is a collection of hazardous waste databases that are accessed through a computer bulletin board. The bulletin board includes features such as news items, bulletins, and special interest conferences including the Bioremediation Special Interest Group. It also features a message board that enables users to share ideas and questions. The central component of ATTIC is the ATTIC Database, which contains abstracts and summaries from technical documents and reports that are both keyword and free-text searchable. ATTIC is staffed by the Scientific Consulting Group, Inc.

Subject emphasis: alternative treatment clean-up technology for hazardous waste.

Internet via-web: http://www.epa.gov/access/chapter3/s1-13.html

E-mail: internet_support@unixmail.rtpnc.gov

Phone: (301) 670-6294 (System Operator); (301) 670-3808 (Online Computer Access)

Fax: (301) 670-3815

Mailing address: ATTIC/Technical Support/SCG, Inc., 4 Research Place, Suite 210, Rockville, MD 20850

EPA Asbestos Ombudsman Clearinghouse/Hotline

Description: The assigned mission of the Asbestos Ombudsman Clearinghouse/Hotline is to provide to the public sector, including individual citizens and community services, information on handling abatement and management of asbestos in schools, the work place, and the home. Interpretation of the asbestos-in-schools requirements is provided. Publications to explain recent legislation are also available. The Asbestos Hazard Emergency Response Act (AHERA) of 1986 assigned duties of the EPA Asbestos Ombudsman to include asbestos abatement/management

Internet via-web: http://www.epa.gov/access/chapter3/s3-1.html

E-mail: internet_support@unixmail.rtpnc.gov

Phone: (703) 305-5938; (800) 368-5888

Fax: (703) 305-6462

Mailing address: U.S. Environmental Protection Agency, Asbestos Ombudsman, 1230C, 401 M Street, SW, Washington, DC 20460

EPA Clean Lakes Clearinghouse

Description: The Clean Lakes Clearinghouse is a resource center for information on lake and watershed restoration, protection, and management. The Clearinghouse includes a bibliographic database of Clean Lakes Program Reports, conference proceedings, technical reports, government documents, journal articles, and books. Data can be accessed by topic, keyword, state or country, EPA region, waterbody name, author, and date. The Terrene Institute staff maintains the database and assists users. Searches can be ordered at a cost of $50 per hour, plus postage and handling.

Subject emphasis: restoration, management, and protection of lakes.

Internet via-web: http://www.epa.gov/access/chapter3/s3-14.html

E-mail: internet_support@unixmail.rtpnc.gov

Phone: (202) 833-8317; (800) 726-LAKE

Fax: (202) 296-4071

Mailing address: Clean Lakes Clearinghouse, The Terrene Institute, 1717 K Street, NW, Suite 801, Washington, DC 20006

EPA Clean-Up Information Bulletin Board System

Description:	The CLU-IN Bulletin Board offers a number of services including online messages and bulletins, computer files, programs, and databases, and special interest group areas. Messages and bulletins may be read online while longer computer files, databases, and models may be either uploaded (sent) or downloaded (received) through CLU-IN. The system targets those involved in hazardous waste remediation and corrective action activities, and is intended to provide an efficient mechanism for the exchange of technological information. The universe of users includes EPA Headquarters, regional and laboratory staff, state and local officials, contractors, consultants, academic institutions, private organizations, and the public. Subject emphasis: hazardous waste site clean-up technologies, activities.
Internet via-web:	http://www.epa.gov/access/chapter3/s1-14.html
E-mail:	internet_support@unixmail.rtpnc.gov
Phone:	(301) 589-8366 (System Access); (301) 589-8368 (System Operator); or (703) 308-8827 (Project Officer)
Fax:	(301) 589-8487
Mailing address:	U.S. Environmental Protection Agency, Technology Innovation Office, 5102W, 401 M Street, SW, Washington, DC 20460

EPA Clearinghouses and Hotlines

Description: A collection of all US EPA public information clearinghouses, hotlines, and electronic bulletin boards. The information has been developed by the Environmental Protection Agency (EPA) to respond to legislative initiatives requiring the Agency to provide outreach, communications, and technology transfer. The information can be found through the following steps: go to http://www.epa.gov/, select Finding EPA Information-Libraries, Hotlines, Information Locators, select Clearinghouses and Hotlines.

Clearinghouses facilitate the networking and exchange of critical information. Many clearinghouses use bulletin boards and hotlines to provide convenient access for remote users. Clearinghouses are also useful as a central access point for hard-to-locate technical reports and documents.

With the exception of The National Response Center, all clearinghouses and hotlines listed in this chapter are maintained by EPA and its agents.

Internet via-web: http://www.epa.gov/

E-mail: internet_support@unixmail.rtpnc.gov

EPA Control Technology Center

Description: The CTC provides technical support and guidance on air pollution emissions and control technology, as well as general information on the Federal Small Business Assistance Program. Service includes Hotline: direct, quick access to EPA experts; Engineering Assistance: short term, detailed assistance to resolve source specific issues; Technical Guidance: CTC documents, computer software, and workshops. The CTC also supports the RACT/BACT/LAER Clearinghouse and provides access to the Global Greenhouse Gases Technology Transfer Center.

Subject emphasis: air emissions and air pollution control technology for all air pollutants including air toxics emitted by stationary sources, and information on the Federal Small Business Assistance Program, RACT/BACT/LAER Clearinghouse, and Global Greenhouse Gases Technology Transfer Center.

Internet via-web: http://www.epa.gov/access/chapter3/s1-3.html

E-mail: internet_support@unixmail.rtpnc.gov

Phone: (919) 541-0800

Fax: (919) 541-0072

Mailing address: U.S. Environmental Protection Agency, Emission Standards Division, Office of Air Quality Planning and Standards, MD-13, Research Triangle Park, NC 27711

EPA Emergency Planning and Community Right-to-Know Information Hotline

Description: The EPA's Emergency Planning and Community Right-to-Know Information Hotline's primary function is to provide regulatory, policy, and technical assistance to federal agencies, local and state governments, the public, the regulated community, and other interested parties in response to questions related to the Emergency Planning and Community Right-to-Know Act (Title III of SARA). The Hotline provides information on the availability of documents related to Title III of SARA and provides copies of selected documents related to Title III of SARA on a limited basis.

Subject emphasis: Emergency Planning and Community Right-to-Know Act (Title III of the Superfund Amendments and Reauthorization Act (SARA)).

Internet via-web: http://www.epa.gov/access/chapter3/s1-15.html

E-mail: internet_support@unixmail.rtpnc.gov

Phone: (703) 412-9877; (800) 535-0202; (800) 553-7672 (TDD)

Fax: (703) 412-3333

Mailing address: Booz Allen & Hamilton, Inc., 1725 Jefferson Davis Highway, Arlington, VA 22202

EPA Emission Factor Clearinghouse

Description:	The Clearinghouse is a means of exchanging information on air pollution control matters, between and among federal, state and local pollution control agencies, private citizens, universities, contractors, and foreign governments. It addresses the criteria pollutants (Particulate/PM-10, Total Organic Compounds, SO2, NOx, CO, and Lead) and toxic substances from stationary and area sources, as well as mobile sources. Subject emphasis: air pollutant emission factors and estimation tools for criteria and toxic pollutants from stationary and area sources, as well as mobile sources.
Internet via-web:	http://www.epa.gov/access/chapter3/s1-4.html
E-mail:	internet_support@unixmail.rtpnc.gov
Phone:	(919) 541-5477
Fax:	(919) 541-0684
Mailing address:	U.S. Environmental Protection Agency, Emission Inventory Branch, MD-14, Research Triangle Park, NC 27711

EPA Environmental Financing Information Network

Description: EFIN provides an online database with publication abstracts and referrals to a network of public financing and environmental program experts. Help with database and literature searches is available, upon request. The EFIN database is accessed directly through several electronic information systems--Public Technology, Inc. (PTI) Local Exchange (LEX), National Conference of State Legislatures (NCSL) LEGISNET, and Government Finance Officers Association (GFOA) Government Finance Network (GF-NET). Call these organizations for EFIN access directions: PTI/LEX (202) 626-2400, NCSL/LEGISNET (303) 830-2200, GFOA/GF-NET (1-800) 829-GFNT. EPA staff access EFIN through the NCSL.

Subject emphasis: financing alternatives for state and local environmental programs and projects (for example, public drinking water, wastewater treatment, and solid waste infrastructure). Information on alternative financing mechanisms for environmental protection, state revolving funds and public-private partnerships is included.

Internet via-web: http://www.epa.gov/access/chapter3/s3-23.html

E-mail: internet_support@unixmail.rtpnc.gov

Phone: (202) 260-0420

Fax: (202) 260-0710

Mailing address: U.S. Environmental Protection Agency, EFIN, 3304, 401 M Street, SW, NELC 014, Washington, DC 20460

EPA Green Lights Program

Description: The Green Lights Program provides information on energy efficient lighting and how companies can join and become a partner or ally with the Green Lights Program. An EPA speaker travels around the United States encouraging companies to join. Eight hundred companies have joined, and 13 states have agreed to convert all state and government buildings to energy efficient lighting in the next 5 years.

Subject emphasis: energy efficient lighting.

Internet via-web: http://www.epa.gov/access/chapter3/s1-6.html

E-mail: internet_support@unixmail.rtpnc.gov

Phone: (202) 775-6650

Fax: (202) 775-6680

Mailing address: The Bruce Company, 1850 K Street, NW, Suite 290, Washington, DC 20006

EPA Hazardous Waste Ombudsman Program

Description: The hazardous waste management program established under RCRA is the most complex regulatory program developed by EPA. It assists the public and regulated community in resolving problems concerning any program or requirement under the Hazardous Waste Program. The Ombudsman Program, located at Headquarters and in each Regional office, handles complaints from citizens and the regulated community, obtains facts, sorts information, and substantiates policy.

Subject emphasis: Resource Conservation and Recovery Act (RCRA).

Internet via-web: http://www.epa.gov/access/chapter3/s1-16.html

E-mail: internet_support@unixmail.rtpnc.gov

Phone: (202) 260-9361; (800) 262-7937

Fax: (202) 260-8929

Mailing address: U.S. Environmental Protection Agency, Hazardous Waste Ombudsman Program, 5101, 401 M Street, SW, Room SE 315, Washington, DC 20460

EPA Indoor Air Quality Information Clearinghouse

Description: Opened in Fall 1992, IAQ INFO provides access to a full range of information about indoor air quality problems. The clearinghouse is equipped with toll-free, operator-assisted telephone access, and is able to provide written information including fact sheets and brochures, perform literature searches, and make referrals to appropriate government and not-for-profit organizations.

Subject emphasis: indoor air quality.

Internet via-web: http://www.epa.gov/access/chapter3/s1-7.html

E-mail: internet_support@unixmail.rtpnc.gov

Phone: (301) 585-9020; (800) 438-4318

Fax: (301) 588-3408

Mailing address: Indoor Air Quality Information Clearinghouse, PO Box 37133, Washington, DC 20013-7133

EPA INFOTERRA

Description: INFOTERRA is an international environmental referral and research service made up of 155 countries coordinated by the United Nations Environment Programme (UNEP) in Nairobi, Kenya. The mission of the INFOTERRA network is to link national and international institutions and experts in a cooperative venture to improve the quality of environmental decisionmaking worldwide.

The U.S. National Focal Point for the INFOTERRA network is located at EPA Headquarters. The services of INFOTERRA/USA include responding to international requests for environmental information through document delivery, database searching, bibliographic products, and referrals to experts. Additionally, INFOTERRA/USA responds to international requests received by EPA staff, and will assist U.S. residents in identifying sources for international environmental information.

Subject emphasis: international environmental information.

Internet via-web: http://www.epa.gov/access/chapter3/s2-1.html

E-mail: internet_support@unixmail.rtpnc.gov

Phone: (202) 260-5917

Fax: (202) 260-6257

Mailing address: U.S. Environmental Protection Agency, INFOTERRA/USA National Focal Point, 3404, 401 M Street, SW, M 2904, Washington, DC 20460

EPA Institute

Description: As the national clearinghouse for all Agency training activities, from environmental services to enforcement to personal and professional development, the Institute focuses on in-house training, but serves as the Agency's training "broker" with other agencies. The EPA Institute, in addition, is a trainer of trainers, consultant to training course designers, and registrar for all Institute approved courses.

Subject emphasis: Institute training.

Internet via-web: http://www.epa.gov/access/chapter3/s3-24.html

E-mail: internet_support@unixmail.rtpnc.gov

Phone: (202) 260-3351

Fax: (202) 260-6786

Mailing address: U.S. Environmental Protection Agency, EPA Institute, 3632, 401 M Street, SW, Room 3241, Washington, DC 20460

EPA International Cleaner Production Information Clearinghouse

Description: The International Cleaner Production Information Clearinghouse (ICPIC) is located in the Paris office of the United Nations Environment Programme (UNEP), Industry and Environmental Programme Activity Center, and implemented under UNEP's Cleaner Production Programme. ICPIC and its sister system PIES share several common databases, such as the Case Studies and General Publications databases, and communicate each day to update files and relay messages. ICPIC contains a message center, bulletin board, calendar of events, case studies, descriptions of national and international programs, bibliography of clean technology documents, and a directory of contacts. ICPIC may be accessed directly or through PIES.

Subject emphasis: pollution prevention, source reduction, recycling, and substitution.

Internet via-web: http://www.epa.gov/access/chapter3/s3-9.html

E-mail: internet_support@unixmail.rtpnc.gov

Phone: 33 1 40 58 88 50 (France); (703) 821-4800 (USA); System Telephone: 33 1 40 58 88 78 (France); (703) 506-102 (USA)

Fax: 33 1 40 58 88 74 (France); (703) 821-4775 (USA)

Mailing address: UNEP IE/PAC, OzonAction Programme, Tour Mirabeau, 39-43,quai Andre Citroen, 75739 Paris Cedex 16 France; or PPIC/OzonAction, c/o SAIC, 7600-A Leesburg Pike, Falls Church, VA 22043 USA

EPA Methods Information Communications Exchange

Description: The methods section implemented 'MICE' in 1991 to better handle incoming technical questions or comments on its "Test Methods for Evaluating Solid Waste--Physical/Chemical Methods" (SW-846). Questions regarding the Toxicity Characteristic Leaching Procedure (TCLP), organic analyses, inorganic analyses, miscellaneous tests, and quality control are answered by chemists, ground water specialists, sampling experts and other professionals who are experienced and knowledgeable in SW-846 testing procedures. The MICE is integrated to an answering machine which is available 24 hours a day. Callers are instructed to leave their name, affiliation, phone number, and question or comment. The messages are retrieved on a daily basis. The questions are researched and the phone calls are returned within one business day.

Subject emphasis: analytical test methods (SW-846) for the characterization of hazardous waste in support of Resource Conservation and Recovery Act (RCRA).

Internet via-web: http://www.epa.gov/access/chapter3/s1-17.html

E-mail: internet_support@unixmail.rtpnc.gov

Phone: (703) 821-4789

Fax: (703) 821-4719

Mailing address: Methods Information Communications Exchange, Falls Church, VA 22043

EPA Model Clearinghouse

Description: Established at the request of the EPA Regional Offices, the EPA Model Clearinghouse reviews dispersion modeling techniques for criteria pollutants in specific regulatory applications. Public access to historical Agency decisions concerning deviations from modeling guidelines, as well as periodic reports published by the Clearinghouse, can be accomplished through PC computer hookup [(919) 541-5742] to the Support Center for Regulatory Air Models, Bulletin Board System (SCRAM BBS).

Subject emphasis: interpretation of modeling guidance.

Internet via-web: http://www.epa.gov/access/chapter3/s1-5.html

E-mail: internet_support@unixmail.rtpnc.gov

Phone: (919) 541-5683

Mailing address: U.S. Environmental Protection Agency, Office of Air Quality Planning and Standards, Source Receptor Analysis Branch, MD-14, Research Triangle Park, NC 27711

EPA National Air Toxics Information Clearinghouse

Description: The primary purpose of the National Air Toxics Information Clearinghouse is to collect, classify, and disseminate air toxics (noncriteria pollutant) information submitted by state and local air agencies, and to make the audience aware of published air toxics information from EPA, other federal agencies, and similar relevant sources. State and local information includes general Agency facts, regulatory program descriptions, acceptable ambient limits, permitted facilities, source testing data, emissions inventories, and monitoring.

Subject emphasis: air toxics (noncriteria air pollutants) and the development of air toxics control programs.

Internet via-web: http://www.epa.gov/access/chapter3/s1-8.html

E-mail: internet_support@unixmail.rtpnc.gov

Phone: (919) 541-0850

Fax: (919) 541-4028

Mailing address: U.S. Environmental Protection Agency, Office of Air Quality Planning and Standards, MD-13, Research Triangle Park, NC 27711

EPA National Lead Information Center Hotline

Description: The mission of the National Lead Information Center Hotline is to provide information to parents on how to protect children from lead poisoning. Callers hear a recording--in either English or Spanish--that requests their name and address. Each person who leaves his or her name and address will receive an information package consisting of a brochure on how to protect children from lead poisoning, three related fact sheets, and a list of state and local contacts who can provide additional information. The brochure and fact sheets are available in English or Spanish.

Subject emphasis: information to help parents protect their children from lead poisoning in the home; list of state and local contacts.

Internet via-web: http://www.epa.gov/access/chapter3/s3-2.html

E-mail: internet_support@unixmail.rtpnc.gov

Phone: (800) 532-3394; TDD number for hearing-impaired: (800) 526-5456

EPA National Pesticide Information Retrieval System

Description: NPIRS, a subscription database of the Center for Environmental and Regulatory Systems (CERIS), provides information (for subscribers only) on pesticide products (current and historical) that have been registered by the EPA. Registration support documents, commodity/tolerance data, Material Safety Data Sheets, Fact Sheets, and state product registration data are provided.

Subject emphasis: EPA pesticide product registration information with focus on agriculture.

Internet via-web: http://www.epa.gov/access/chapter3/s3-3.html

E-mail: internet_support@unixmail.rtpnc.gov

Phone: (317) 494-6614

Fax: (317) 494-9727

Mailing address: CERIS (NPIRS), 1231 Cumberland Avenue, Suite A, West Lafayette, IN 47906-1317

EPA National Pesticide Telecommunications Network (NPTN)

Description: NPTN, managed by Texas Tech University Health Sciences Center, Lubbock, Texas, is a free service providing a variety of impartial information concerning pesticides; pesticide product information; information on recognition and management of pesticide poisonings; toxicology and symptomatic reviews; referrals for laboratory analyses, investigation of pesticide incidents, and emergency treatment information; safety information; health and environmental effects; and cleanup and disposal procedures.

Subject emphasis: pesticides.

Internet via-web: http://www.epa.gov/access/chapter3/s3-4.html

E-mail: internet_support@unixmail.rtpnc.gov

Phone: (800) 858-7378 (General public); (800) 858-7377 (Medical and government personnel)

Fax: (806) 743-3094

Mailing address: National Pesticide Telecommunications Network

EPA National Radon Hotline

Description: A message records names and addresses of callers, and a brochure on radon is sent via first class mail. The Hotline is used for consumer research to estimate the impact of EPA advertising and outreach to targeted, at-risk populations.

Subject emphasis: information on radon health effects and testing homes for radon. Radon information callers will receive a brochure and a coupon for a discount radon test kit.

Internet via-web: http://www.epa.gov/access/chapter3/s1-9.html

E-mail: internet_support@unixmail.rtpnc.gov

Phone: (800) SOS-RADON, 24 hour toll free hotline

Fax: (202) 293-0032

Mailing address: National Safety Council, Environmental Health Center, Suite 401, 1019 19th Street, NW, Washington, DC 20036

National Radon Hotline, Box 33435, Washington, DC 20035-0435

EPA National Response Center

Description: The National Response Center receives reports of oil, hazardous chemical, biological, and radiological releases. The NRC then passes those reports to a predesignated federal On-Scene Coordinator (OSC), who coordinates cleanup efforts, and other responsible federal agencies.

Subject emphasis: oil, hazardous chemical, biological, and radiological releases.

Internet via-web: http://www.epa.gov/access/chapter3/s1-18.html

E-mail: internet_support@unixmail.rtpnc.gov

Phone: (202) 267-2675, (800) 424-8802

Fax: (202) 267-2181

Mailing address: National Response Center, U.S. Coast Guard Headquarters, 2100 Second Street, SW, Room 2611, Washington, DC 20593-0001

EPA National Small Flows Clearinghouse

Description: The purpose of the Clearinghouse is to collect, classify, and disseminate information on alternative wastewater technology to assist small communities in wastewater management. The Clearinghouse distributes publications (general information, technical manuals, brochures, and case studies) and videotapes, performs literature searches, operates a toll-free hotline, produces free newsletters, and operates a computer bulletin board.

Subject emphasis: small community wastewater treatment.

Internet via-web: http://www.epa.gov/access/chapter3/s3-15.html

E-mail: internet_support@unixmail.rtpnc.gov

Phone: (304) 293-4191, (800) 624-8301

Mailing address: National Small Flows Clearinghouse, West Virginia University, PO Box 6064, Morgantown, WV 26506-6064

EPA Nonpoint Source Information Exchange

Description: NPS Information Exchange publishes the bulletin NPS News-Notes 8 times a year. Target audience is state and local water quality managers (and other interested public officials, environmental groups, private industry, citizens, and academics). Circulation is over 10,000.

NPS Electronic Bulletin Board System (NPS BBS) is a 24-hour telecommunications system that provides current information to an audience similar to that of News-Notes. It is used as a means to exchange text and program computer files, as an information resource, and as a forum for open discussion. Several "mini-bulletin boards" accessed through the main board allow parties with specialized interests to share information. Some of these Special Interest Group (SIG) Forums are Agriculture, Waterbody System Support, NPS Research, and Fish Consumption Advisories. Also on-line are searchable databases such as the Clean Lakes Clearinghouse, NPS News-Notes database, Water Quality Educational Materials Index, and Fish Consumption Bans and Advisories database.

Whenever possible, the NPS Information Exchange will direct requests for NPS information and documents to appropriate sources within EPA.

Subject emphasis: nonpoint source water pollution and other water environment-related issues.

Internet via-web: http://www.epa.gov/access/chapter3/s3-16.html

E-mail: internet_support@unixmail.rtpnc.gov

Phone: (202) 260-3665

Fax: (202) 260-1517

Mailing address: U.S. Environmental Protection Agency, Office of Wetlands, Oceans, and Watersheds, Assessment and Watershed Protection, Division, 4503, 401 M Street, SW, Washington, DC 20460

EPA Office of Air Quality Planning and Standards Technology Transfer Network Bulletin Board System

Description: The OAQPS TTN is a network of electronic bulletin boards that provides information and technology exchange in different areas of air pollution control ranging from emission test methods to regulatory air quality models. The purpose of the boards is to foster technology transfer among all parties interested in the solution of the nation's air pollution problems.

Subject emphasis: air pollution and related topics.

Internet via-web: http://www.epa.gov/access/chapter3/s1-10.html

E-mail: internet_support@unixmail.rtpnc.gov

Phone: (919) 541-5742; (919) 541-5384

Mailing address: Office of Air Quality Planning and Standards, Technology Transfer Network Bulletin Board System, Research Triangle Park, NC 27711

EPA Office of Research and Development Electronic Bulletin Board System

Description: The ORD BBS is an online, text-searchable database of every ORD publication produced since 1976 (more than 18,526 citations). Each citation includes title, abstract, ordering information, and much more. The ORD BBS also offers such features as messages, bulletins of new information, public domain files, online registration for ORD meetings, and currently has seven specialty areas, such as water, regional operations, expert systems, biotechnology, Quality Assurance/Quality Control (QA/QC), Environmental Monitoring and Assessment Plan, and Integrated Risk Information System.

Subject emphasis: communication and technology transfer.

Internet via-web: http://www.epa.gov/access/chapter3/s3-13.html

E-mail: internet_support@unixmail.rtpnc.gov

Phone: (513) 569-7610; (800) 258-9605

Fax: (513) 569-7566

Mailing address: U.S. Environmental Protection Agency, Center for Environmental Research Information, 26 West Martin Luther King Drive, Cincinnati, OH 45268

EPA Office of Water Resource Center

Description:	The Resource Center distributes documents produced by the Office of Ground Water and Drinking Water, the Office of Science and Technology, and the Municipal Support Division of the Office of Wastewater Enforcement and Compliance. Subject emphasis: ground water and drinking water documents, water quality criteria documents and videotapes, and wastewater treatment documents.
Internet via-web:	http://www.epa.gov/access/chapter3/s3-17.html
E-mail:	internet_support@unixmail.rtpnc.gov
Phone:	(202) 260-7786
Fax:	(202) 260-4383
Mailing address:	Office of Water Resource Center, 401 M Street, SW, 4100 (P), Washington, DC 20460

EPA OzonAction

Description: OzonAction is designed to provide national and international programmatic and technical information on alternatives to ozone depleting substances (ODS) identified for phaseout under the Montreal Protocol. Data and information are provided on five industry use sectors: solvents, coatings and adhesives, aerosols, foams, halons, and refrigeration and air-conditioning. OzonAction contains technology case studies, a database of ODS-reduction products and services, national and corporate programme summaries, experts, literature database of significant ODS reduction documents, and message centers. OzonAction also relays the solvent substitute database known as OZONET, compiled by the Industry Cooperative for Ozone Layer Protection (ICOLP). OzonAction will be used nationally as the electronic network arm of the Office of Air and Radiation's Stratospheric Ozone Information Hotline to assist in conveying ODS information and EPA-approved alternatives as required under Title VI of the Clean Air Act.

Subject emphasis: ozone-depleting substance (ODS) alternatives; pollution prevention; and source reduction, recycling, and substitution.

Internet via-web: http://www.epa.gov/access/chapter3/s3-10.html

E-mail: internet_support@unixmail.rtpnc.gov

Phone: 33 1 40 58 88 50 (France); (703) 821-4800 (USA); System Telephone: 33 1 40 58 88 78 (France); (703) 506-102 (USA)

Fax: 33 1 40 58 88 74 (France); (703) 821-4775 (USA)

Mailing address: UNEP IE/PAC, OzonAction Programme, Tour Mirabeau, 39-43, quai Andre Citroen, 75739 Paris Cedex 16 France; or PPIC/OzonAction, c/o SAIC, 7600-A Leesburg Pike, Falls Church, VA 22043 USA

EPA Pollution Prevention Information Clearinghouse (PPIC)

Description: The Pollution Prevention Information Clearinghouse (PPIC) is dedicated to reducing or eliminating industrial pollutants through technology transfer, education, and public awareness. It is a free, nonregulatory service of the U.S. EPA. A Reference and Referral Telephone Service is available to answer questions, take orders for documents distributed by PPIC, or refer callers to appropriate contacts. The Clearinghouse distributes selected EPA documents and factsheets on pollution prevention free of charge. A pollution prevention collection is being established at the U.S. EPA Headquarters Library and will include pollution prevention training materials, conference proceedings, journals, and federal and state government publications. It can be accessed at the Headquarters Library, Room M2904, 401 M Street, SW, Washington, DC.

Subject emphasis: pollution prevention, source reduction, and recycling.

Internet via-web: http://www.epa.gov/access/chapter3/s3-11.html

E-mail: internet_support@unixmail.rtpnc.gov

Phone: (202) 260-1023

Fax: (202) 260-0178

Mailing address: U.S. Environmental Protection Agency, 3404, 401 M Street, SW, Washington, DC 20460

EPA Pollution Prevention Information Exchange System (PIES)

Description:	PIES is the computerized information network of PPIC. It provides online interactive access to a wide range of pollution prevention information. The network is open 24 hours a day and requires no user fees. PIES features literature search functions, a national calendar of conferences and workshops, case studies, a message center, and direct access to news and documents. The International Cleaner Production Information Clearinghouse (ICPIC) and OzonAction are also available through PIES. PIES can be accessed through a personal computer with a modem and communications software. It is accessible through a regular telephone call, the SprintNet network, and the EPA x.25 wide area network (for EPA employees only). Subject emphasis: pollution prevention, source reduction, recycling.
Internet via-web:	http://www.epa.gov/access/chapter3/s3-12.html
E-mail:	internet_support@unixmail.rtpnc.gov
Phone:	(703) 821-4800
Fax:	(703) 821-4775
Mailing address:	Science Applications International Corp., 7600-A Leesburg Pike, Room 369, Falls Church, VA 22043

EPA Reasonably Available Control Technology, Best Available Control Technology, and Lowest Achievable Emission Rate Clearinghouse

Description:	RACT/BACT/LAER Clearinghouse provides state and local air pollution control agencies, EPA regional offices, and other interested parties with current and historical information on control technology determinations. These determinations relate to emission controls for existing sources of nonattainment pollutants (RACT) and new source review permits for major new or modified sources (BACT and LAER) required under the Clean Air Act Determinations and are made on a case-by-case basis.
Subject emphasis:	Air Pollution Control Technology related to New Source Review Permitting Requirements.
Internet via-web:	http://www.epa.gov/access/chapter3/s1-11.html
E-mail:	internet_support@unixmail.rtpnc.gov
Phone:	(919) 541-0800
Fax:	(919) 541-0072
Mailing address:	U.S. Environmental Protection Agency, Office of Air Quality Planning and Standards, Emissions Standards Division, MD-13, Research Triangle Park, NC 27711

EPA Resource Conservation and Recovery Act/Superfund/Underground Storage Tank Hotline

Description: The Environmental Protection Agency's RCRA/SF/OUST Hotline's primary function is to provide assistance to the public and regulated community and other interested parties in understanding EPA's regulations pursuant to RCRA, UST, CERCLA, and Pollution Prevention/Waste Minimization. In addition to providing regulatory support, the Hotline also provides information on RCRA, UST, CERCLA, and Waste Minimization/Pollution Prevention documents.

Subject emphasis: RCRA, Underground Storage Tanks (UST), Superfund/CERCLA, and Pollution, Prevention/Waste Minimization.

Internet via-web: http://www.epa.gov/access/chapter3/s1-19.html
E-mail: internet_support@unixmail.rtpnc.gov
Phone: (703) 412-9810; (800) 424-9346; (800) 553-7672 (TDD)
Fax: (703) 412-3333
Mailing address: RCRA/SF/OUST Hotline, 1725 Jefferson Davis Highway, Arlington, VA 22202

EPA Risk Communication Hotline

Description: The Risk Communication Hotline responds to questions from EPA program and regional offices, the academic/scientific community, and the general public. The hotline provides information on EPA's Risk Communication Program, responds to questions on risk communication issues and literature, and makes referrals to other related agency sources of information. A small library is maintained with a bibliographic database, reports, articles, books, conference summaries, risk communication materials, and training materials. Available publications are mailed out upon request.

Subject emphasis: risk communication.

Internet via-web: http://www.epa.gov/access/chapter3/s3-28.html
E-mail: internet_support@unixmail.rtpnc.gov
Phone: (202) 260-5606
Fax: (202) 260-9757
Mailing address: U.S. Environmental Protection Agency, Risk Communication Hotline, 2127, Room 425, West Tower, 401 M Street, SW, Washington, DC 20460

EPA Safe Drinking Water Hotline

Description:	The Safe Drinking Water (SDW) Hotline assists both the regulated community (public water systems) and the public with their understanding of the regulations and programs developed in response to the Safe Drinking Water Act Amendments of 1986. Subject emphasis: Safe Drinking Water Act and Amendments.
Internet via-web:	http://www.epa.gov/access/chapter3/s3-18.html
E-mail:	internet_support@unixmail.rtpnc.gov
Phone:	(800) 426-4791
Mailing address:	U.S. Environmental Protection Agency, Office of Ground Water and Drinking Water, 4601, Resource Center, 401 M Street, SW, Washington, DC 20460

EPA Solid Waste Assistance Program

Description: To help increase the availability of information in the field of municipal solid waste management, the USEPA Office of Solid Waste has funded SWAP, the Solid Waste Assistance Program. SWAP is a technical information service designed to collect and distribute materials upon individual request. Information available through SWAP is extensive, and is intended to provide assistance to government agencies, professional associations, industry, citizen groups, and other interested parties on all aspects of solid waste management. Peer Match is an assistance program that aids state and local governments by connecting knowledgeable municipal solid waste professionals, with communities in need of municipal solid waste assistance. Established by the Solid Waste Association of North America (SWANA), and funded through the United States Environmental Protection Agency (USEPA), the program matches public sector municipal solid waste experts with public sector professionals in need of assistance. The Peer Match Program will coordinate the match and can pay up to one-half the associated travel expenses, with the requesting organization covering the other half. Dial the technical assistance line at 1-800-677-0297 for more information. Areas of concentration are bio-medical waste, collection, composting, household hazardous waste, landfill gas recovery, landfilling, materials processing, planning, recycling, regionalization, siting, source reduction, transfer, waste-to-energy.

Subject emphasis: all aspects of solid waste management, including source reduction, recycling, composting, planning, education and training, public participation, legislation and regulation, waste combustion, collection, transfer, disposal, landfill gas, and special wastes.

Internet via-web: http://www.epa.gov/access/chapter3/s1-20.html

E-mail: internet_support@unixmail.rtpnc.gov

Phone: (800) 677-9424

Fax: (301) 585-0297

Mailing address: Solid Waste Assistance Program, PO Box 7219, 8750 Georgia Avenue, Silver Spring, MD 20907

EPA Stratospheric Ozone Information Hotline

Description: The Stratospheric Ozone Information Hotline offers consultation on ozone protection regulations and requirements under Title VI of the Clean Air Act Amendments (CAAA) of 1990. Title VI covers the following key aspects of the production, use, and safe disposal of ozone-depleting chemicals: 1) production phaseout and controls, 2) servicing of motor vehicle air conditioners, 3) recycling and emission reduction, 4) technician and equipment certification, 5) approval of alternatives, 6) a ban of nonessential uses, 7) product labeling, and 8) federal procurement. The hotline is a distribution center and referral point for information on other general aspects of stratospheric ozone depletion and its protection. The hotline maintains a library of relevant policy and science documents, reports, articles, and contact lists.

Subject emphasis: stratospheric ozone depletion and protection.

Internet via-web: http://www.epa.gov/access/chapter3/s1-12.html

E-mail: internet_support@unixmail.rtpnc.gov

Phone: (800) 296-1996

Fax: (202) 783-1106

Mailing address: c/o The Bruce Co., 501 3rd Street, NW, Washington, DC 20001

EPA The 33/50 Program, Special Projects Office (SPO), Office of Pollution Prevention and Toxics

Description: The Environmental Protection Agency's 33/50 Program is a voluntary initiative seeking commitments from companies to reduce their reported releases and off-site transfers of 17 targeted chemicals. The program provides a new framework for environmental stewardship based on voluntary, cooperative efforts among local communities, industry, and government.

The overall goal of the 33/50 Program is to promote the benefits of pollution prevention while obtaining measurable reductions in pollution. The 33/50 Program is named after its goals of 33 percent reduction by the end of 1992 and 50 percent reduction by the end of 1995.

If a company wishes to participate in the program, it should send a short letter to EPA stating a corporatewide numerical reduction goal for the sum of the 17 target chemicals for 1995. This numerical commitment is the sole requirement for participation in the 33/50 Program.

Internet via-web: http://www.epa.gov/access/chapter3/s3-5.html

E-mail: internet_support@unixmail.rtpnc.gov

Phone: (202) 260-6907

Fax: (202) 260-1764

Mailing address: The 33/50 Program, U.S. Environmental Protection Agency, (Mail Code 7408), 401 M Street, SW, Washington, DC 20460

EPA Toxic Release Inventory User Support

Description: Congress mandated that U.S. manufacturers (SIC codes 20-39) must report to the EPA annually on their releases to the environment of about 330 toxic chemicals. The Toxic Release Inventory User Support (TRI-US) was established by the Office of Pollution Prevention and Toxics in September, 1990 to facilitate access to TRI data. TRI-US provides access and user support to EPA staff, other federal agencies, industry, environmental and public interest groups, libraries, the international community and citizens. The Toxic Release Inventory is available in a variety of formats: online, CD ROM, microfiche, floppy diskettes, and hard copy. TRI-US provides help in choosing and accessing all formats with documentation and specialized search assistance. Customized online searches are performed on a limited basis for patrons of TRI-US. Training and demonstrations of the TRI online system and the CD ROM are available through TRI-US. General questions about the Toxic Pesticides and Toxic Substances Release Inventory are answered, and referrals to EPA Regional or State TRI contacts, libraries, or other TRI resource centers are provided.

A special Emergency Planning and Community Right-to-Know Act (EPCRA) collection and a vertical file of TRI materials are located in the Office of Pollution Prevention and Toxics (OPPT) Library and continue to grow with additions of articles, studies, and reports. TRI data products are maintained and available in the library.

TRI-US provides literature searches and current information updates to EPA staff and state TRI contacts on a regular basis and on demand. TRI-US facilitates providing new products to EPA staff and the states and public access information products to the public.

Subject emphasis: Toxic Release Inventory/Emergency Planning and Community Right-to-Know Act, public access, training.

Internet via-web: http://www.epa.gov/access/chapter3/s3-6.html

E-mail: internet_support@unixmail.rtpnc.gov

Phone: (202) 260-1531

Fax: (202) 260-4659

Mailing address: U.S. Environmental Protection Agency, TRI User Support, 7407, 401 M Street, SW, Washington, DC 20460

EPA Toxic Substances Control Act Assistance Information Service

Description: The TSCA Assistance Information Service provides information on TSCA regulations. Technical as well as general information is available.

Subject emphasis: Toxic Substances Control Act (TSCA) regulatory information.

Internet via-web: http://www.epa.gov/access/chapter3/s3-7.html

E-mail: internet_support@unixmail.rtpnc.gov

Phone: (202) 554-1404

Fax: (202) 554-5603

Mailing address: U.S. Environmental Protection Agency, Environmental Assistance Division, 7408, 401 M Street, SW, Washington, DC 20460

EPA Wastewater Treatment Information Exchange

Description: The computer bulletin board maintained at the National Small Flows Clearinghouse, called the Wastewater Treatment Information Exchange Bulletin Board Service (WTIE-BBS), provides a direct forum for discussion of ideas and exchange of information about small-scale wastewater systems. WTIE-BBS is a free service accessible anywhere in the United States on a 24-hour basis. It allows users to "post" questions and notices, to converse with others, and to download information. WTIE-BBS services include electronic mail, electronic conferencing, surveys, and news bulletins.

Subject emphasis: small community wastewater programs.

Internet via-web: http://www.epa.gov/access/chapter3/s3-20.html

E-mail: internet_support@unixmail.rtpnc.gov

Phone: (800) 544-1936

Fax: (304) 293-3161

Mailing address: Wastewater Treatment Information Exchange, National Small Flows Clearinghouse, West Virginia University, PO Box 6064, Morgantown, WV 26506-6064

EPA Wastewi$e Program

Description: This program is a voluntary partnership with EPA, businesses, designed to reduce municipal solid waste. The focus is on (1) waste prevention, (2) recycling collection, and (3) buying or manufacturing recycled products.

Internet via-web: http://cygnus-group.com/ULS/Waste/epa.html

Environmental Recycling Hotline

Description: A uniform resource locator (URL) directory that hosts a variety of server lists related to information about recycling programs. The information is organized by zip code.

Internet via-web: http://www.primenet.com/erh.html

E-mail: lippard@primenet.com

Phone: (800) 94-reuse

Environmental Sensors

Description: A supplier of a range of instruments, equipment, and systems services for scientific, agricultural, and environmental applications. Information includes moisture point, smart sensors, virrib, oilspy, optic, and weather sentry.

Internet via-web: http://www.envsens.com/

E-mail: jjohnsto@envsens.com

Phone: (800) 553-3818; (619) 486-5688

Fax: (619) 486-1899

Environmental Technologies USA, Inc.

Description: A leader in environmentally sound plastics. It owns two subsidiaries: Clean Green Packing ("Clean Green") and United Recycling, Inc. ("URI"). Clean Green holds proprietary patents and license rights to manufacture and sell biodegradable/water soluble, starch-based loose-fill packing materials (replaces polystyrene foam packing peanuts). URI holds proprietary technologies for converting used consumer carpeting primarily into nylon or nylon/polypropylene material for manufacturing into commercial and consumer products. URI also recently announced a strategic corporate relationship with Fluor Daniel, Inc.

Internet via-web: http://www.mvisibility.com/ENVR/

E-mail: envr@mvisibility.com

Phone: (212) 661-3848

Mailing address: 325 Cedar Avenue South, Minneapolis, MN 55454

For an investor package call or write: Market Visibility, Inc., 605 Third Avenue, New York, NY 10158; Ph: 800-814-0569; Fax: 212-867-0524

Environmentally Conscious Design and Manufacturing Lab

Description: A uniform resource locator (URL) directory that hosts a variety of server lists related to Environmentally Conscious Design and Manufacturing. Maintained by the University of Windsor.

Internet via-web: http:/ie.uwindsor.ca/ecdm_lab.html

Mailing address: University of Windsor

Enviroene

Description: A uniform resource locator (URL) directory that hosts a variety of server lists related to pollution prevention and waste minimization. The program, funded by the Environmental Protection Agency and the Strategic Environmental Research and Development Program, allows those implementing pollution prevention programs or developing research and development projects to benefit from the experience, progress, and knowledge of their peers. Enviroene includes a pollution prevention forum for all levels of government, researchers, industry, and public interest groups.

Internet via-web: http://wastenot.inel.gov/envirosense/

ETS International, Inc.

Description: An analytical company that specializes in toxic emission measurement and control as well as infrastructure design, construction, and maintenance.

Internet via-web: http://www.infi.net/~etsas/

Phone: (540) 265-0004

Mailing address: 1401 Municipal Road, NW, Roanoke, VA 24012

Evaluation of Artificial Wetlands

Description: *Evaluation of Artificial Wetlands for Filtering of Agricultural Waste Waters* is an annotated Bibliography under the Canada-Nova Scotia Agreement on the Agricultural Component of the Green Plan: Sustainable Agri-Food Waste Management Program: Project GP-022.

Internet via-web: http://gus.nsac.ns.ca/~piinfo/resman/wetlands/anno/annobib.html

The Exchange

Description: This program was created by a grant from the U.S. EPA and now funded and overseen by Earthcycle, the National Materials Exchange Network has been the free local and international on-line marketplace for trading and recycling used and surplus materials and goods for over six years.

With over 10,000 current listings (including 3,000 outside the U.S.), the Network allows state waste exchanges and industrial and consumer users at no charge to immediately list or search among thirty categories of available and wanted items as diverse as scrap and precious metals, paper products, surplus chemicals, computer components, gems, textiles, auto, and boat equipment and electronics goods.

Internet via-web: http://www.earthcycle.com/g/p/earthcycle//

Exfluor Research Corp.

Description: A manufacturer of various chemicals. Specific chemicals of interest can be found by searching through the Internet homepage, WWW Chemicals, at the Internet address, "http://www3.ios.com:80/~ilyak."

Internet via-web: http://www3.ios.com:80/~ilyak/ind008.html

Phone: (512) 454-3812

Fax: (512) 453-5016

Mailing address: 8868 Research Blvd. #206, Austin, TX 78758-6446

Falcon Software

Description: A publisher of an educational software. Their business specializes in chemistry, engineering, electronics, environmental science.

Internet via-web: http://www.falconweb/index.html

E-mail: falconinfo@falconsoftware.com

Phone: (617) 235-1767

Fax: (617) 235-7026

Mailing address: One Hollis Street, Wellesley, MA 02181

Farrell Research

Description: An analytical company that specializes in X-ray florescence for lead in paint and soil, environmental information, environmental products and software, and federal regulations.

Internet via-web: http://www.ionet.net/~enviro/enviro.shtml

E-mail: enviro@ionet.net

Phone: (918) 252-5068

Fax: (918) 254-0218

Mailing address: 856 S Aspen #18, Broken Arrow, OK 74012

Fast Heat

Description: A manufacturer of runnerless molding systems, electronic temperature controls, and electric heating elements for plastics industry.

Internet via-web: http://www.ais.net/fastheat/

E-mail: fasheat@fastheat.com

Phone: (708) 833-5400

Fax: (708) 833-2040

Mailing address: Elmhurst, Illinois

FedWorld Information Network

Description: The National Technical Information Service (NTIS) introduced FedWorld in November 1992 to help with the challenge of accessing U.S. government information online. In an electronic age when more and more government agencies are racing to get online, FedWorld can help you keep up with the flood of information worldwide. NTIS is providing a comprehensive central access point for locating and acquiring government information.

The goal of NTIS FedWorld is to provide a one-stop location for the public to locate, order and have delivered to them, U.S. government information.

Internet via-web: http://www.fedworld.gov/index.html

E-mail: webmaster@fedworld.gov

Phone: (703) 487-4850

Mailing address: 5285 Port Royal Road, Springfield, VA 22161

Fiberglass World

Description: A trade association that promotes trade in the Fiberglass Industry. Includes fiberglass associations, publications, industry directory, recycling, calendar of events,

Internet via-web: http://www.fiberglass.com/fiberglass/index.html

E-mail: webmaster@fiberglass.com

Finishing Industry Home Page

Description: An information clearinghouse for surface finishing industry. The surface finishing industry includes anodizing, conversion coating, electroforming, electroplating, electroless plating, electropolishing, galvanizing, metal finishing, painting, phosphatizing, powder coating, printed circuits, and the allied surface finishing arts.

Internet via-web: http://www.finishing.com/

E-mail: tmooney@intac.com

Fisher Scientific

Description: A manufacturer of various chemicals. Fisher's phone directory includes customer service: (800) 766-700, fax (800) 926-1166, laboratory product help: (800) 388-8355, safety product help: (800) 926-8999, instrument service: (800) 395-5442.

Internet via-web: http://www.fisher1.com/

E-mail: cservice@fisher1.com

Phone: (800) 766-7000

Fax: (800) 926-1166

Mailing address: 711 Forbes Ave., Pittsburg, PA 15219-4785

FMC Corporation

Description: Producer of five broad markets: industrial chemicals, performance chemicals, precious metals, defense systems, and machinery and equipment. FMC operates 99 manufacturing facilities and mines in 21 countries.

Internet via-web: http://fmcweb.ncsa.uiuc.edu/home.html

Mailing address: 200 East Randolph Drive, Chicago, IL 60601

Food and Drug Administration (FDA)

Description: FDA is an agency within the Public Health Service, which in turn is a part of the Department of Health and Human Services. FDA is a public health agency, charged with protecting American consumers by enforcing the Federal Food, Drug, and Cosmetic Act and several related public health laws. It is FDA's job to see that the food we eat is safe and wholesome, the cosmetics we use won't hurt us, the medicines and medical devices we use are safe and effective, and that radiation-emitting products such as microwave ovens won't do us harm. Feed and drugs for pets and farm animals also come under FDA scrutiny. FDA also ensures that all of these products are labeled truthfully with the information that people need to use them properly.

Internet via-web: http://www.fda.gov

Internet via-telnet: fdabbs.fda.gov; login: bbs; password: bbs

Phone: (301) 443-4908

Ford Motor Co. Recycling

Description: Ford's programs on the recycling of plastics, radiators, batteries, etc., from used cars.

Internet via-web: http://www.ford.com/corporate-info/environment/Recycling.html

Foundation for Cross-Connection Control and Hydraulic Research

Description: A research institute that is maintained by the University of Southern California. The institute offers training courses and training tools to assist those who are involved in cross-connection control.

Internet via-web: http://www.usc.edu:80/dept/fccchr/

E-mail: fccchr@usc.edu

Phone: (213) 740-2032

Fax: (213) 740-8399

Mailing address: University of Southern California, KAP-200 University Park MC-2531

The Frank E. VanLare Wastewater Treatment Facility

Description: A municipal wastewater treatment facility. The facility is located in Rochester, New York, USA. Provides general information, history, and diagrams.

Internet via-web: http://www.history.rochester.edu/class/vanlare/home.htm

Mailing address: Rochester, New York

FRC International

Description: A recycling company that reclaims Halon 1211 and Halon 1301 found in fire extinguishers and fire suppression systems, and purchases systems that are being taken out of service.

Internet via-web: http://norden1.com/~frc/

E-mail: frc@nordenl.com

Friends of Earth Home Page

Description: A uniform resource locator (URL) directory that hosts a variety of servers in environmental areas. The servers provided include Newbury bypass campaign; the chemical release inventory.

Internet via-web: http://www.foe.co.uk

E-mail: webmaster@foe.co.uk

Furuuchi Chemical Co.

Description: A manufacturer of many chemical products including high purity metals, high purity metal compounds, superconductor, superconductive oxide materials, superconductive CO-Precipitation powder, semiconductors, advanced ceramics, cleaning solutions for semiconductor, SEMICO CLEAN23, SEMICO CLEAN56, vacuum coating materials, electron beam tablets, sputtering targets, thin film materials, fluoride glass, fluoride glass fibers, single crystals, laboratory reagent, laboratory wares, consultant.

Internet via-web: http://www.bekkoame.or.jp/~kittel/

Fax: 81-3-3766-8310

Mailing address: Tokyo Tatemono Bldg. 2-4-18 Ohmori-kita, Ota-ku, Tokyo 143 Japan

Garden State Laboratories

Description: A certified independent environmental and food testing laboratory. The company analyzes drinking water, wastewater, sludges, biosolids, soils, solid wastes, hazardous wastes, monitoring wells, food and dairy products for chemical, bacteriological and other microbiological contaminants.

Internet via-web: http://www.planet.net/gsl/

Phone: (800) 273-8901

Fax: (908) 688-8966

Mailing address: 410 Hillside Avenue, Hillside, NJ 07205

Gas Processors Association (GPA)

Description: A professional organization of operating and producing companies engaged in the processing of natural gas. GPA today is an incorporated, nonprofit trade association made up of approximately 165 corporate members, all of whom are engaged in the processing of natural gas into a merchantable pipeline gas, or in the manufacture, transportation, or further processing of liquid products from natural gas.

The active membership as a group account for approximately 90 percent of all natural gas liquids produced in the United States. The active membership also includes a number of Canadian and foreign companies that produce natural gas liquids on a global scale.

Internet via-web: http://www.galstar.com/~gpa/

Gelman Sciences

Description: A developer and manufacturer of microporous membranes and devices for critical microfiltration requirements. With manufacturing operations across the United States, they have a worldwide distribution and service network.

Internet via-web: http://argus-inc.com/Gelman/Gelman.html

E-mail: info@gelman.com

Phone: (800) 521-1520; (313) 913-6197

Fax: (800) 926-4588

Genentech, Inc.

Description: A biotechnology company that discovers, develops, manufactures, and markets human pharmaceuticals for medical needs.

Internet via-web: http://outcast.gene.com/

E-mail: webmaster@gene.com

Phone: (415) 225-2309

Fax: (415-225) 1411

Mailing address: 460 Pt. San Bruno Blvd., So., San Francisco, CA 94080

General Electric (GE) Plastics

Description: A manufacturer of engineering plastics. GE plastics offers a spectrum of basic resin chemistries: lexan polycarbonate resin, cycolac ABS resin, noryl modified polyphenylene oxide-based resin, prevex phenylene ether resin, valox TBT resin, supec polyphenylene sulfide resin, geloy ASA resin, etc.

Internet via-web: http://www.ge.com/gep/homepage.html

E-mail: plastics@www.ge.com

Phone: (413) 448-7110

Mailing address: One Plastics Ave, Pittsfield, MA 01201

General Electric WWW Server

Description: A diversified technology, manufacturing and services company operating on a worldwide basis. Their business includes aircraft engines, capital services, electrical distribution and control, information services, lighting, NBC, plastics, power systems, research, and development.

Internet via-web: http://www.ge.com/

E-mail: geinfo@www.ge.com

Phone: (800) 626-2004, (518) 438-6500

General Plastex

Description: A plastics engineering company that offers customers a wide range of products and services including conversion of vented to nonvented barrels, complete honing service, remanufactured screws, upsizing and downsizing of injection molding machines for greater efficiency and best use of shot size, total design and manufacturing for new screws.

Internet via-web: http://www.polysort.com/gplastex

Phone: (800) 777-4719; (216) 745-7775

Fax: (216) 745-6939

Mailing address: 1050 Eagon Street, Barberton, OH 44203

Generic Bulletin Board Builder (GENBBB)

Description: A uniform resource locator (URL) directory that hosts a variety of servers related to chemical and process engineering. The servers provided include ASPEN PLUS virtual library, Edinburgh University Chemical Engineering Department, Process Improvement laboratory, etc.

Internet via-web http://www.cs.colorado.edu/homes/mcbryan/public_html/bb/167/summary.html

Geraghty & Miller, Inc.

Description: A provider of integrated investigation, engineering, and construction services to their clients through a network of offices. Known particularly for providing groundwater solutions, Geraghty & Miller is a full-service firm capable of investigating and remediating environmental problems in all media.

Internet via-web: http://www.gmgw.com/gm

Mailing address: Netherlands

Gilchrist Polymer Center

Description: A plastics engineering company that specializes in services to the polymer industry.

Internet via-web: http://www.polysort.com/gilchrist

Phone: (216) 733-2400

Fax: (216) 733-0663

Mailing address: 3375 Gilchrist Road, Mogadore, OH 44260

Gilson, Inc.

Description: A supplier of HPLC autoinjectors. Features of the injector include five built-in injection methods, three loop filling options, manu-driven software, wide selection of racks and vials.

Internet via-web: http://pubs.acs.org/pin/gilson/gil222p1.html

Phone: (800) 445-7661

Mailing address: 3000 W. Beltline Highway, Box 620027, Middleton, WI 53562-0027

Global Network of Environment and Technology (GNET)

Description: A consulting firm that provides constantly updated information on innovative environmental technologies, business, and news, with leads to marketing intelligence, financing, and contracting opportunities. GNET seeks to promote sustainable development and environmental remediation through technological innovation, with a focus upon commercializing Department of Energy developed applications. Features moderated discussion forums as well as full-text search capabilities.

Internet via-web: http://www.gnet.org/

E-mail: gnet@gnet.org

Global Recycling Network, Inc.

Description: A uniform resource locator (URL) directory that hosts a variety of servers related to recycling and reuse industry.

Internet via-web: http://grn.com/grn/

E-mail: grn@grn.com;

Phone: (516) 286-5580

Fax: (516) 286-5551

Mailing address: 2715A Montauk Hwy, Brookhaven, New York 11719

Goodyear Tire & Rubber Company

Description: A manufacturer of tires, rubbers, and chemicals. Also provides the care and feeding of your tires, the tire wear advisory, the buyers guide, the calendar of events, and related linkages.

Internet via-web: http://www.goodyear.com/

E-mail: webmaster@goodyear.com

Government Printing Office (GPO)

Description: GPO prints, binds, and distributes the publications of the Congress as well as the executive departments and establishments of the federal government. Distribution is being accomplished on an increasing basis via various electronic media in accordance with Public Law 103-40, "The Government Printing Office Electronic Information Access Enhancement Act of 1993." The GPO began operations in accordance with Congressional Joint Resolution 25 of June 23, 1860. The activities of the GPO are outlined and defined in the act of October 22, 1968, as amended (44 U.S.C. 101 et sec.).

Internet via-web: http://www.access.gpo.gov

E-mail: wwwadmin@www.access.gpo.gov

Green Design Initiative

Description: A campus-wide Green Design Initiative to promote environmentally conscious engineering, product and process design, manufacturing, and architecture. The initiative involves forming partnerships with industrial corporations, foundations, and government agencies to develop joint research and education programs that improve environmental quality while encouraging sustainable economic development. Maintained by the Carnegie Mellon University.

Internet via-web: http://www.ce.cmu.edu:8000/GDI/

E-mail: dsde+@andrew.cmu.edu

Mailing address: Carnegie Mellon University

Green Engineering

Description: A uniform resource locator (URL) directory that hosts a variety of servers in the area of environmentally conscious activities. The servers provided include Environmentally Conscious Design and Manufacturing Laboratory, National Key Centre for Design, Green Design Initiative, etc.

Internet via-web: http://ie.uwindsor.ca/other_green.html

E-mail: tomw@ie.uwindsor.ca

Green Market

Description: A consulting firm that is in partnership with EcoNet and the Institute for Global Communications. The mission is to bring together progressive and ecologically conscious companies and organizations to provide information and products easily and affordably to concerned, earthwise consumers.

Internet via-web: http://www.igc.apc.org/GreenMarket/

E-mail: greenadmin@greenmarket.com

Greenfield Environmental

Description:	A hazardous materials management company. Information includes people at Greenfield environmental, map of Greenfield facilities, household hazardous waste programs, specific service offering, etc.
Internet via-web:	http://www.greenfield.com/
E-mail:	infosys@a2.greenfield.com
Phone:	(800) 878-9555; (619) 673-6000
Fax:	(619) 673-6015
Mailing address:	15151 Innovation Drive, San Diego, CA 92128

GreenSoft Corporation

Description:	A software company. The purpose is to develop an environmentally friendly software delivery and support system utilizing the Internet and other on-line resources.
Internet via-web:	http://www.greendesk.com/

Greenspan Technology

Description:	An environmental monitoring company that specializes in water quality monitoring technology and services for the water resources, environmental, and pollution markets.
Internet via-web:	http:/peg.pegasus.oz.au/~greenspan/
E-mail:	greenspan@peg.apc.org
Phone:	61-76-61-7699
Fax:	61-76-61-9190
Mailing address:	24 Palmerin Street, Warwick Queensland 4370 Australia

Guide to Chemical Engineering

Description:	A uniform resource locator (URL) directory that hosts a variety of servers for chemists. The servers provided include visualization in chemistry and rendering techniques, chemistry at Rensselaer, Cornell university, etc.
Internet via-web:	http://www.theworld.com/science/engineer/chemical/subject.htm

H&S Chemical Co.

Description: A manufacturer of various chemicals. Specific chemicals of interest can be found by searching through the Internet homepage, WWW Chemicals, at the Internet address, "http://www3.ios.com:80/~ilyak."

Internet via-web: http://www3.ios.com:80/~ilyak/ind020.html

Phone: (513) 841-2424

Fax: (514) 242-4849

Mailing address: 300 Murray Road, PO Box 17186, Cincinnati, OH 45217

HAAKE

Description: A plastics engineering company that specializes in the fields of rheology and laboratory scale polymer processing equipment.

Internet via-web: http://www.polysort.com/haake

Phone: (800) 631-1369

Mailing address: 53 W. Century Rd, Paramus, NJ 07652

Hampford Research, Inc.

Description: A manufacturer of various chemicals. Specific chemicals of interest can be found by searching through the Internet homepage, WWW Chemicals, at the Internet address, "http://www3.ios.com:80/~ilyak."

Internet via-web: http://www3.ios.com:80/~ilyak/ind001.html

Phone: (203) 375-1137

Fax: (203) 386-9754

Mailing address: PO Box 1073, 292 Longbrook Ave., Stratford, CT 06497

Hampshire Chemical Co.

Description: A manufacturer of various chemicals. Specific chemicals of interest can be found by searching through the Internet homepage, WWW Chemicals, at the Internet address, "http://www3.ios.com:80/~ilyak."

Internet via-web: http://www3.ios.com:80/~ilyak/ind003.html

Phone: (617) 861-9700

Fax: (617) 863-8930

Mailing address: 55 Hayden Ave., Lexington, MA 02173

Harbec Plastics

Description: A plastics engineering company that provides a variety of services for the plastics industry.

Internet via-web: http://www.harbec.com/

Mailing address: 369 Route 104, Ontario, NY 14519

Hazardous Materials Management

Description: A Canadian trade magazine. Also provides on-line information on 1996 buyer's guide and directory on hazardous waste management.

Internet via-web: http://www.io.org/~hzmatmg/

E-mail: hazmatmg@inforamp.net

Phone: (905) 305-6155

Fax: (905) 305-6255

Mailing address: 951 Dension Street, Unit 4, Markham, ON L3R 3W9

Hewlett-Packard Analytical

Description: A supplier of measurement, computing, and communication coupled with analytical chemistry.

Internet via-web: http://www.hp.com/go/analytical

E-mail: webmaster@www.hp.com

Phone: (800) 227-9770

Hitachi Instruments, Inc.

Description: A manufacturer of analytical instruments. Information includes product information, service and support information, latest press releases, and links to web homepages.

Internet via-web: http://www.hii.hitachi.com/

E-mail: info@hii.hitachi.com

Phone: (800) 548-9001

Huls America Inc.

Description: A manufacturer of various chemicals. Specific chemicals of interest can be found by searching through the Internet homepage, WWW Chemicals, at the Internet address, "http://www3.ios.com:80/~ilyak."

Internet via-web: http://www3.ios.com:80/~ilyak/ind029.html

Phone: (908) 980-6940

Fax: (908) 980-6970

Mailing address: 80 Centennial Ave., PO Box 456, Piscataway, NJ 08855-04456

Husky Injection Molding Systems

Description: A plastics engineering company that provides a variety of services for the plastics industry.

Internet via-web: http://www.husky.on.ca/

E-mail: webmaster@husky.on.ca

Hydrocomp, Inc.

Description: A consulting firm specializing in hydrologic modeling and analysis.
Internet via-web: http://www.hydrocomp.com/
E-mail: info@hydrocomp.com
Phone: (415) 637-9060
Fax: (415) 637-9976

Hydromantis

Description: A Canadian consulting firm. The company develops and applies computer-based technologies to satisfy the needs and requirements of customers concerned with planning, design, operation, automatic control, and training.
Internet via-web: http://www.hydromantis.com/index.html
E-mail: info@hydromantis.com
Phone: (905) 522-0012
Fax: (905) 522-0031
Mailing address: 1685 Main Street West, Suite 302, Hamilton, Ontario L8S 1G5, Canada

Hypercube, Inc.

Description: A computer software company specializing in molecular modeling analysis. Their products include HyperChem, ChemPlus, and new standards for ease of use and molecular modeling power on PC-based systems.
Internet via-web: http://www.hyper.com/
E-mail: info@hyper.com
Phone: (519) 725-4040
Fax: (519) 725-5193
Mailing address: 419 Phillip Street, Waterloo, Ontario Canada, N2L3X2

IBM World Wide Web

Description: Hosts a variety of IBM web servers around the globe.
Internet via-web: http://www.ibm.com/
E-mail: askibm@info.ibm.com
Phone: (800) IBM-3333

ICI Fiberite

Description: A supplier of advanced composite materials and thermoset molding compounds. Their products range from aircraft and automobiles to transformer bushings and welding equipment.
Internet via-web: http://www.olworld.com./olworld/mall/mall_us/c_busfin/m_fiberi/
Phone: (507) 452-8044

Idetec, S.A. de C.V.

Description: A developer of chemical intermediates and the technologies for pharmacopeia products.
Internet via-web: http://www.spin.com.mx/grupocs/gcs-ideb.html

Imperiali Lab

Description: A research institute. Their research interest focuses on the bioorganic chemistry of amino acids, peptides, and proteins. Has adopted a multidisciplinary approach for the investigation of enzyme-catalyzed protein modification reactions and the de novo design of proteins with unique structural and functional properties.
Internet via-web: http://impind.caltech.edu/
E-mail: ranabir@impind.caltech.edu
Phone: (818) 395-8397
Fax: (818) 564-9297
Mailing address: Division of Chemistry and Chemical Engineering, California Institute of Technology, Pasadena, CA 91125

Indigo Instruments

Description:	An instrumentation manufacturing company that offers a variety of products including instruments and scientific glassware.
Internet via-web:	http://ds.internic.net/indigo/index.html
E-mail:	slogan@indigo.com
Phone:	(519) 746-4761
Fax:	(519) 747-5636
Mailing address:	167 Lexington Court, Unit 6, Waterloo, Ontario, N2J 4R9, Canada

Indofine Chemical Co.

Description:	A manufacturer of various chemicals. Specific chemicals of interest can be found by searching through the Internet homepage, WWW Chemicals, at the Internet address, "http://www3.ios.com:80/~ilyak."
Internet via-web:	http://www3.ios.com:80/~ilyak/INDOFINE.html
Phone:	(908) 356-6778
Fax:	(908) 359-1179
Mailing address:	PO Box 473, Somerville, NJ 08876

Indoor Air Quality (IAQ) Publications

Description:	A publications company that provides resources for news, information, and conferences addressing indoor pollution, indoor air quality, healthy buildings, and lead poisoning prevention.
Internet via-web:	http://www.iaqpubs.com/
E-mail:	iaqpubs@aol.com
Phone:	(800) 394-0115
Fax:	(301) 913-0119
Mailing address:	2 Wisconsin Circle, Suite 430, Chevy Chase, MD 20815

Industrial Plastics and Paints

Description: A manufacturer of plastics and paints. Information includes plastics, paints, etc.
Internet via-web: http://storefront.net/storefront/ipp/index.html
Phone: (800) 667-1757; (604) 727-3545
Mailing address: 3944 Quadra Street, Victoria, B.C. V8X 1J6

Industrial Services International

Description: A manufacturer of superabsorbent polymers. The polymers are solid, water swellable, cross-linked, polymers designed to absorb water or aqueous solutions. The product swells and forms a tight gel that holds water molecules, even under pressure.
Internet via-web: http://www.cais.com/cytex/isi/tsorb.html
E-mail: absorbu@aol.com
Phone: (800) 227-6728; (941) 753-1310
Fax: (941) 758-1175
Mailing address: 4301 32nd St. W, Suite A11-B11, Bradenton, FL 34205

IndustryLink

Description: A uniform resource locator (URL) directory that hosts a variety of servers who are interested in any industrial activities. It links, by industrial grouping: (1) automation, (2) chemicals, (3) energy, (4) environmental, (5) heavy equipment and machinery, (6) manufacturing, (7) mining, (8) pulp and paper, (9) plastics and polymers, (10) cross-industry products and services, (11) miscellaneous industries, (12) all industrial homepages listed alphabetically.
Internet via-web: http://www.industrylink.com/
E-mail: ilink@industrylink.com
Phone: (905) 677-4894
Fax: (905) 677-3421
Mailing address: 5935 Airport Road, Suite 120, Mississauga, Ontario, L4V 1W5 Canada

IndustryNet

Description: A uniform resource locator (URL) directory that hosts a variety of servers in the area of (1) online market place; (2) online service; (3) business and industry on the World Wide Web; (4) maintenance, repair, operation, production (MROP).

Internet via-web: http://www.industry.net/

Info-Labview Mailing List

Description: The info-labview mailing list is used by LabVIEW users to communicate with other users and with National Instruments. To add your name to the list, send an Internet message with your name and Internet address to info-labview-request@pica.army.mil.

Internet via-web: ftp://ftp.natinst.com/README

Information Technology for Environmentally Conscious Design, Construction, and Manufacturing

Description: A uniform resource locator (URL) directory that hosts a variety of server lists related to Environmentally Conscious Design and Manufacturing. Maintained by the Tufts University.

Internet via-web: http://iv.cee.tufts.edu:8000/berger_chair.html

E-mail: wrodrigu@pearl.tufts.edu

Phone: (617) 627-3035

Mailing address: W. Rodriguez, Tufts University, 223 Anderson Hall, CEE/CADS, Medford, MA 02155

Ingot Metal Company, Ltd.

Description: A manufacturer of federalloy. Federalloy alloys are lead-free casting alloys that retain the pressure tightness, castability and associated mechanical properties inherent in the leaded bronze alloys.

Internet via-web: http://www.io.org/~dshore/lead.html

E-mail: dshore@io.org

Phone: (800) 567-7774; (416) 749-1372

Fax: (416) 749-1371

Mailing address: 111 Fenmar Drive, Weston, Ontario, M9L 1M3

Inktomi

Description: An information search engine. Subjects that can be searched in this engine cover most human activities including arts, business, computers, education, engineering, entertainment, government, health, medicine, news, recreation, science, social science, society, and culture.

Internet via-web: http://inktomi.berkeley.edu/

Institute for Gas Utilization and Processing Technologies (IGUPT)

Description: An interdisciplinary research unit at the University of Oklahoma. Their objectives are to develop technologies that will enhance the value and use of natural gas resources.

Internet via-web: http://www.uoknor.edu/igupt/

E-mail: igupt@uoknor.edu

Phone: (405) 325-5813

Fax: (405) 325-5813

Mailing address: 100 East Boyd Ave., Rm. T335, Norman, OK 73019

Integrated Design Engineering Systems (IDES)

Description: A plastics engineering company that provides information systems used to improve and expedite engineering design in the plastics industry.

Internet via-web: http://www.idesinc.com/

Mailing address: 209 Grand Ave, Laramie, Wyoming

Integrating Environment and Development

Description: A uniform resource locator (URL) directory that hosts a variety of server lists related to the National Strategy on Ecologically Sustainable Development in Australia.

Internet via-web: http://www.erin.gov.au/portfolio/esd/integ.html

Intera Inc.

Description:	A multi-disciplinary environmental consulting firm serving business, industry, government, and fellow members of the environmental profession. Applies technology to data analysis, data interpretation, and solution-oriented decision making.
Internet via-web:	http://www.nmia.com/~interags/home/home.html
E-mail:	interages@nmia.com
Mailing address:	headquartered in Austin, Texas

Interactive Simulations, Inc.

Description:	A molecular modeling software vendor specializing in solving critical scientific problems for the biotechnology and pharmaceutical industries, primarily through the development of innovative software for molecular modeling, simulation, and drug design.
Internet via-web:	http://www.intsim.com/~isigen/
E-mail:	sales@intsim.com
Phone:	(619) 658-9462
Fax:	(619) 658-9463
Mailing address:	5330 Carrol Canyon Road, Suite 203, San Diego, CA 92121

Interchem Corporation

Description:	A supplier of (1) bulk pharmaceuticals, (2) fine chemical intermediates, and (3) custom manufacturing.
Internet via-web:	http://www.interchem.com/
E-mail:	webmaster@interchem.com
Mailing address:	120 Rt. 17 North, Paramus, NJ 07652

Interduct

Description:	A uniform resource locator (URL) directory that hosts a variety of server lists related to green engineering in the Netherlands.
Internet via-web:	http://dutw239.tudelft.nl
E-mail:	vanWijk@interduct
Mailing address:	Deflt University Clean Technology Institute in the Netherlands

International Organization for Standardization (ISO)

Description: ISO is a worldwide federation of national standards bodies from some 100 countries, one from each country. ISO, headquartered in Switzerland, is a nongovernmental organization established in 1947. The mission of ISO is to promote the development of standardization and related activities in the world with a view to facilitating the international exchange of goods and services, and to developing cooperation in the spheres of intellectual, scientific, technological, and economic activity.

Internet via-web: http:/www.iso.ch/

E-mail: central@isocs.iso.ch

Phone: 41-22-749-0111

Fax: 31-22-733-3430

Mailing address: 1, rue de Varembe, Case postale 56, CH-1211 geneve 20, Switzerland

International Society of Heterocyclic Chemistry

Description: The ISHC now has a joint World Wide Web home page with the Royal Society Heterocyclic Group.

Internet via-web: http://euch6f.chem.emory.edu/ishc.html

Internet Chemistry Resources

Description: A uniform resource locator (URL) directory that hosts a variety of servers in the selected chemistry and associated fields. Resources are organized according to either Internet service (gopher, ftp, etc.) or subject (e.g., teaching resources).

Internet via-web: http://www.rpi.edu/dept/chem/cheminfo/chemres.html

E-mail: wardej@rpi.edu

Phone: (518) 276-6000

Mailing address: Rensselaer Polytechnic Institute, 110 8th Street, Troy, NY 12180

Internet Chemistry Resources

Description: A uniform resource locator (URL) directory that hosts a variety of servers.

Internet via-web: http://euch6f.chem.emory.edu/ishc.html

ISO 14000 Information

Description: The emerging set of ISO 14000 standards is the most comprehensive environmental quality management initiative ever undertaken. The first edition of the standard is expected to be issued in 1996 and will prescribe requirements for environmental quality management as defined by the International Organization for Standardization in Geneva, Switzerland.

The ISO 14000 standard will support the corporate goals of achieving compliance with legal requirements, establishing internal environmental quality policies, and managing marketplace expectations. These goals will be accomplished by implementing environmental quality management systems, environmental audits, environmental performance evaluations, product life-cycle assessments, and product labeling.

The ISO 14000 series of standards emerged primarily as a result of two events: the Uruguay round of the GATT negotiations begun in 1986 and the Rio Conference on the Environment held in 1992. The GATT talks addressed the need to avoid or remove non-tariff barriers to trade, while the Rio Conference established the world's commitment to protection of the environment. The ISO 14000 series standards represent a new consensus position for business and the environmental community: They are a blueprint for promoting world trade while encouraging and assisting organizations to be environmentally responsible. It is no longer a question of jobs or the environment, but the standard will now allow jobs and trade growth while promoting a clean environment.

Internet via-web: http://www.stoller.com/iso.htm

E-mail: iso14000@stoller.com

Phone: (303) 546-4373

IVAM Environmental Research

Description: A research institute that is maintained by the University of Amsterdam. Addresses product studies and life cycle analysis, production process studies, energy studies, company environmental management, sustainable building, international studies, and computer studies.

Internet via-web: http://www.ivambv.uva.nl/welcome.html

E-mail: webmaster@ivambv.uva.nl

Mailing address: University of Amsterdam

J.M. Huber Corporation

Description: A diversified natural resource and industrial product company. Their business focuses on (1) electronics, energy, engineered minerals, specialty chemicals, and wood products. Their products are in finished goods ranging from toothpaste to tires, from computers to paint, from plastics to paneling.

Internet via-web: http://www.huber.com/

E-mail: webmaster@huber.com

Phone: (908) 549-8600

Fax: (908) 549-2239

Mailing address: 333 Thornall Street, Edison, NJ 08818

Jamestown Tooling & Machining

Description: A plastics engineering company that specializes in 3-D Mold Buildings.

Internet via-web: http://www.jtm.com/

E-mail: Sales@spidernet.jtm.com

Phone: (716) 665-4223

Fax: (716) 665-4225

Mailing address: 1376 East Second Street Jamestown, NY 14701

Jandel Scientific

Description:	A developer of software for scientific analysis. Their products include PC-based software for chemical, biochemical, and chemical engineering research.
Internet via-web:	http://www.jandel.com/
E-mail:	sales@jandel.com
Phone:	(415) 453-6700; (800) 4jandel
Fax:	(415) 453-7769
Mailing address:	San Rafael, CA

JEOL USA, Inc.

Description:	A subsidiary of JEOL, Ltd. of Akishima, Japan. The primary business of JEOL is the manufacture, sale, and service of electronic microscopes, such as SEMs and TEMs, and various types of analytical instruments including mass spectrometers, NMRs.
Internet via-web:	http://www.jeol.com/
E-mail:	webmaster@jeol.com
Phone:	(508) 535-5900
Fax:	(508) 535-7741
Mailing address:	11 Dearborn Road, Peabody, MA 01960

Jost Chemical

Description:	A manufacturer of various chemicals. Specific chemicals of interest can be found by searching through the Internet homepage, WWW Chemicals, at the Internet address, "http://www3.ios.com:80/~ilyak."
Internet via-web:	http://www3.ios.com:80/~ilyak/ind007.html
Phone:	(314) 352-0787
Mailing address:	3260 Brannon, St. Louis, MO 63139

K.R. Anderson Co., Inc

Description: A supplier of specialty manufacturing materials such as silicone, solder, epoxies, and many other hi-tech products.

Internet via-web: http://www.kranderson.com/

E-mail: grichards@kranderson.com

Phone: (800) 538-8712; (408) 727-2800

Mailing address: 2800 Bowers Avenue, Santa Clara, CA 95051

Keith Ceramic Materials

Description: A manufacturer of ceramic materials. Their products include synthetic mullite, refractory bonds, refractory plasticizing agent, and investment casting.

Internet via-web: http://www.ceramics.com/~ceramics/keith/

Phone: +44-181-311-8299

Fax: +44-181-311-8238

Mailing address: Fishers Way, Belvedere, Kent DA176BN, England

Khem Products, Inc.

Description: A software company that develops custom-built regulatory compliance software.

Internet via-web: http://www.khem.com/khem/home.html

E-mail: info@khem.com

Phone: (410) 679-6620

Fax: (410) 679-6625

Mailing address: 1217 Bush Road, PO Box 161, Abingdon, Maryland 21009

Kimberlyte Inc.

Description:	A developer of educational software for Macintosh and Windows. The company provides organic chemistry reactions commonly taught in a first year collegiate organic course. The contents include text descriptions, color molecular structures, and reaction mechanisms.
Internet via-web:	http://hookomo.aloha.net/~mikei/kimbhome.html
E-mail:	mikei@aloha.net
Mailing address:	2326 Armstrong Street, Honolulu, HI 96822

Knight-Ridder Information - Dialog

Description:	A uniform resource locator (URL) directory that hosts a variety of servers for business, research, and scientific professionals worldwide.
Internet via-web:	http://www.dialog.com/
Phone:	(415) 254-7000
Fax:	(415) 254-7070
Mailing address:	2440 El Camino Real, Mountain View, CA 94040

Laboratory Equipment Exchange

Description:	This is a buy, sell, and trade magazine for used scientific equipment. There are no upfront charges to place an ad and it is completely free to read and inquire about any of the advertisements. A 5% transaction fee is charged only if a transaction is made.
Internet via-web:	http://www.magic.mb.ca/~econolab/
E-mail:	econolab@magic.mb.ca

Lackie & Associates

Description: A plastics engineering company that manufactures products from recycled materials.

Internet via-web: http://www.recycle.net/recycle/Trade/rs000811.html

E-mail: lackie@recycle.net

Phone: (519) 621-7569

Fax: (519) 621-6227

Mailing address: 1177 Franklin Blvd., Unit #1
Cambridge, Ontario, N1R 7W4 Canada

Lakewood Systems

Description: A manufacturer of battery-powered data recorders for use with a broad range of environmental and industrial instruments.

Internet via-web: http://cban.worldgate.edmonton.ab.ca/lkwd/

E-mail: lakewood@agt.net

Phone: (403) 450-3867 (Canada); (916) 988-7970 (USA)

Fax: (403) 462-9110 (Canada); (916) 988-7994 (USA)

Mailing address: Canadian Corporate Headquarters: 9258-34A Avenue, Edmonton, AB, Canada T6E 5P4

U.S. Corporate Headquarters: 9477 Greenback Lane, #527, Folsom, CA 95630

Lancaster Synthesis, Inc.

Description: A manufacturer of various chemicals. Specific chemicals of interest can be found by searching through the Internet homepage, WWW Chemicals, at the Internet address, "http://www3.ios.com:80/~ilyak."

Internet via-web: http://www3.ios.com:80/~ilyak/Lancaster.html

E-mail: 71650.426@compuserve.com

Phone: (603) 889-3306; (800) 238-2324

Fax: (603) 889-3326

Mailing address: PO Box 1000, Windham, NH 03087-9777

Lanxide Coated Products

Description: A developer of inorganic composite technologies and the manufacturer of titanium carbide (TiC)-coated graphite. TiC is one of the hardest materials yet discovered. TiC-coated graphite components demonstrate exceptional resistance to wear, corrosion, and oxidation isna wide variety of environments.

Internet via-web: http://www.ravenet.com/lanxcoat/

E-mail: lanxcoat@ravenet.com

Phone: (302) 456-6206

Fax: (302) 454-1712

Mailing address: 1300 Marrows Road, PO Box 6077, Newark, Delaware 19714-6077

Lauren Manufacturing Company

Description: A plastics engineering company that provides original equipment manufacturers and fabricators with a complete line of extruded closed-cell sponge, and extruded and molded dense rubber seals, gaskets, and weather stripping.

Internet via-web: http://www.polysort.com/lauren

Phone: (800) 683-0676; (216) 339-3373

Fax: (216) 339-7166

Mailing address: 2228 Reiser Ave., S.E., New Philadelphia, OH 44663

LCA at the University of Toronto

Description: A research institute that is maintained by the University of Toronto. This focuses on life-cycle analysis (LCA) to measure the effectiveness of pollution prevention.

Internet via-web: http://www.ecf.toronto.edu/~young/

E-mail: young@ecf.toronto.edu

Leco Corporation

Description: A supplier of instruments for (1) analyses of conductive (or non-conductive) materials; (2) nitrogen/protein in foods; (3) image analysis of metals, minerals, composites, and other materials.

Internet via-web: http://pubs.acs.org/pin/leco/lec.html

Library of Congress (LOC)

Description: A government-owned library located in Washington DC.

Internet via-web: http://www.loc.gov

Internet via-gopher: gopher://gopher.loc.gov

E-mail: lcweb@loc.gov

Low Gravity Transport Phenomena Laboratory

Description: A research institute funded by NSF. Their research program includes flow visualization and diffusion in liquids metals, segregation during crystal growth in a low-gravity environment, and oscillatory flow and their effect on separation.

Internet via-web: http://gibbs.che.ufl.edu/lowgravity.shtml

E-mail: postmaster@gibbs.che.ufl.edu

Phone: (904) 492-0862

Mailing address: Department of Chemical Engineering, University of Florida, Gainesville, FL 32611

MacroFAQS

Description: A plastics engineering company that provides complete polymer analytical and research services.

Internet via-web: http://www.polysort.com/FAQS

Materials and Electrochemical Research (MER) Corporation

Description: A developer of technologies including (1) fullerenes and nanotechnology, (2) coatings, (3) reinforcements, (4) electrochemical systems, (5) powders, and (6) joining.

Internet via-web: http://www.opus1.com/~mercorp/index.html

E-mail: mercorp@opus1.com

Phone: (520) 574-1980

Fax: (520) 574-1983

Mailing address: 7960 South Kolb Rd, Tucson, AZ 85706

The Materials Systems Laboratory

A research institute that is maintained by the Massachusetts Institute of Technology (MIT). MIT has pioneered the development of practical methods to assess the "utility," that is the value to the user, of otherwise not comparable measures of the performance of materials (such as their weight, reliability, and price). These techniques enable producers, working closely with consumers, both to anticipate market needs and to improve the performance of their industrial strategy.

Internet via-web: http://web.mit.edu/org/c/ctpid/www/msl/index.html

Mailing address: Massachusetts Institute of Technology, MA

Maxima Plastics

Description: A plastics engineering company that combines the flexibility and ingenuity of a small shop with the most modern equipment available.

Internet via-web: http://www.net-link.net/maxima

Phone: (616) 385-2300

Fax: (616) 385-2390

Mailing address: 5266 Lovers Lane, Kalamazoo, MI 49002

MDL Information Systems, Inc.

Description: A supplier of chemical information management software, databases, and services to the pharmaceutical, agrochemical, and chemical industries.

Internet via-web: http://www.mdli.com/

E-mail: jobs@mdli.com

Fax: (510) 614-3679

Mailing address: 14600 Catalina Street, San Leandro, CA 94577

Melamine Chemicals, Inc.

Description: A producer of melamine. The company is one of two producers of melamine in the western hemisphere and is one of the three largest producers in the world.

Internet via-web: http://www.melamine.com/

E-mail: webmaster@melamine

Phone: (504) 473-0525

Fax: (504) 473-0558

Mailing address: Donaldsonville, LA

The Membrane Technology Group

Description: An academic research institute. It is located in the University of Wente, Netherlands. Their research focuses on membrane development, characterization, and process development.

Internet via-web: http://utct1029.ct.utwente.nl/documents/membrane.html

E-mail: membrane@ct.utwente.nl

Mailing address: PO Box 2177, 7500 AE Enschede, The Netherlands

Merck & Company, Inc.

Description: A provider of pharmaceutical products and services. The company develops, manufactures, and markets products to improve human and animal health.

Internet via-web: http://www.merck.com/

E-mail: inf@Merck.com

Mailing address: Whitehouse Station, NJ

Metcalfe Plastics Corporation

Description:	A plastics engineering company that specializes in the manufacture of automotive accessories.
Internet via-web:	http://www.metcalfe.com
Phone:	(714) 838-5228
Fax:	(714) 838-3183
Mailing address:	14751-B Franklin Ave., Tustin, CA 92680

Methanex

Description:	A manufacturer and marketer of methanol. Methanol is also known as methyl alcohol (CH3OH). It is manufactured from synthesis gas, produced from steam-reformed natural gas and carbon dioxide. Methanol is used as a raw material in the production of formaldehyde, MTBE, acetic acid and numerous other chemical derivatives. Provides glossary of terms related to methanol.
Internet via-web:	http://www.methanex.com/invest/

Microanalytics Instrumentation

Description:	A manufacturer of instruments and systems for multidimensional gas chromatography. The range of company services include instrument design and manufacture, technical support services, and analytical support services.
Internet via-web:	http://www.mdgc.com/
E-mail:	dwight@mdgc.com
Phone:	(512) 218-9873
Fax:	(512) 218-9875
Mailing address:	2713 Sam Bass Road, Round Rock, TX 78681

MicroMath Scientific Software, Inc.

Description:	A developer of scientific software for scientists and engineers. The company's products and services include nonlinear system modeling and fitting, program for solving aqueous chemical equilibrium problems, pharmacokinetic library, chemical kinetic library, diffusion library, and Macintosh desktop utility program.
Internet via-web:	http://www.micromath.com/~mminfo/
Phone:	(800) 942-6284
Fax:	(801) 943-0299
Mailing address:	PO Box 71550, Salt Lake City, UT 84171-0550

Micromeritics Instrument Corp.

Description:	An ISO 9001 certified manufacturer of particle measurement technologies. The company manufactures a broad line of particle technology instrumentation for the evaluation of particle size, surface area, pore size and pore size distribution, material density, surface and catalytic activity, and zeta potential.
Internet via-web:	http://www.micromeritics.com/
E-mail:	webmaster@micromeritics.com
Phone:	(770) 662-3656

Millipore Corporation

Description:	A developer of purification technologies. The application of the technology ranges from bacteria testing of water, to sterilization of biopharmaceutical proteins, to eliminating contamination from gases used in manufacturing the latest and hottest semiconductor device.
Internet via-web:	http://www.millipore.com/
Phone:	(800) 645-5476, (617) 275-9200
Fax:	(617) 275-5550
Mailing address:	80 Ashby Road Bedford, MA 01730-2271

Misco International, Inc.

Description: A manufacturer of quality cleaning and maintenance products for professional, commercial, institutional, and industrial applications. Their products include (1) cleaner/degreasers, (2) food service, (3) drain maintainers, (4) housekeeping, (5) disinfectants, (6) floor care, and (7) carpet care.

Internet via-web: http://www.radiks.net/misco/catalog1/

Phone: (800) 231-0915

Mailing address: Messner, Wheeling, Illinois 60090

Miton Products

Description: A manufacturer of specialty construction chemical sealer products, including driveway sealer.

Internet via-web: http://www.internetpagework.com/miton.html

E-mail: aulicino@direct.ca

Phone: (604) 270-7775

Fax: (604) 270-1274

Mailing address: Unit 3, 2900 Simpson Road, Richmond, BC, Canada V6X 2P9

Mittelhauser

Description: A consulting firm that specializes in environmental consulting, engineering, and remediation. It employs highly skilled individuals from a well-balanced variety of disciplines including engineering, geological, and hydrogeological sciences, environmental, risk assessment, regulatory specialists, and field services personnel.

Internet via-web: http://www.mittelhauser.com

E-mail: info@mittelhauser.com

Mobil Corporation

Description: A major oil, gas, and petrochemical company with operations in more than 100 countries. Their other businesses include mining and land development. Information includes exploration and producing, marketing and refining, mobile chemical, mobile mining and minerals, mobile real estate, and mobile technology company.

Internet via-web: dhttp://www.mobil.com/

Moldflow Ltd

Description: A plastics engineering company that provides design and manufacture of plastic parts.

Internet via-web: http://www.worldserver.pipex.com/moldflow/

Phone: 61 3 720 2088

Fax: 61 3 729 0433

Mailing address: 259-261-Colchester Road, Kilsyth, Victoria 3137, Australia

Molecular Simulations, Inc.

Description: A developer of software for many applications. The company develops and markets innovative scientific software and services designed to improve the efficiency of the research, development, and production processes in both life science and materials industries worldwide.

Internet via-web: http://www.msi.com/

Phone: (617) 229-9800

Fax: (617) 229-9899

Mailing address: 16 New England Executive Park, Burlington, MA 08103-5297

Molten Metal Technology

Description: A developer of elemental recycling processes. Their products include (1) Catalytic Extraction Processing (CEP), a patented Elemental Recycling™ technology for converting hazardous waste to useful materials; (2) Quantum-CEP™ technology for reducing and stabilizing; and (3) recycling radioactive waste.

Internet via-web: http://www.mmt.com/

Phone: (617) 487-9700

Mailing address: 51 Sawyer Road, Waltham, MA 02154

Monsanto Company

Description:	One of the major chemical companies in the United States. Their products include (1) crop protection, (2) lawn and garden products, (3) laminated glass with saflex plastic interlayer, (4) ortho yard basics, (5) rubber testing instruments, (6) therminol heat transfer fluids, (7) triaminononame, and (8) vydyne nylon.
Internet via-web:	http://www.monsanto.com/
E-mail:	use Internet for communication
Phone:	(314) 694-1000
Mailing address:	800 North Lindberg Blvd, St. Louis, MO 63167

Montreal Protocol on Substances That Deplete the Ozone Layer

Description:	This treaty is to protect the ozone layer by taking precautionary measures to control equitably total global emissions of substances that deplete it, with the ultimate objective of their elimination on the basis of developments in scientific knowledge, taking into account technical and economic considerations and bearing in mind the developmental needs of developing countries.
Internet via-web:	http://www.greenpeace.org/~intlaw/mont-htm.html

Morflex, Inc.

Description:	A manufacturer of various chemicals. Specific chemicals of interest can be found by searching through the Internet homepage, WWW Chemicals, at the Internet address, "http://www3.ios.com:80/~ilyak."
Internet via-web:	http://www3.ios.com:80/~ilyak/ind023.html
Phone:	(910) 292-1781
Fax:	(910) 854-4058
Mailing address:	2110 High Point Road, Greensboro, NC 27403

Morris Environmental

Description:	A consulting firm that specializes in oil spill response, support, training, and consulting.
Internet via-web:	http://www.morrisenv.com/
E-mail:	jesse_u@morrisenv.com

Mother Lode Plastics

Description: A plastics engineering company that offers a complete custom plastic injection molding facility that includes engineering, tooling, molding, assembly, and delivery.

Internet via-web: http://www.sonnet.com/webworld/mplastic.htm

E-mail: mlpinc@sonnet.com

Phone: (800) 345-7885; (209) 532-5146

Fax: (209) 532-6312

Mailing address: 20833 Mechanical Dr., Sonora, CA 95370

Multibase

Description: A plastics engineering company that offers custom thermoplastic compounds and services.

Internet via-web: http://www.polysort.com/multibase

Phone: (216) 867-5124

Fax: (216) 666-7419

Mailing address: 3835 Copley Road, Copley, Ohio 44321

National Association of Environmental Professionals (NAEP)

Description: A professional organization concerned with the standards of ethics and competency within the environmental professions.

Internet via-web: http://enfo.com/NAEP/

Phone: (202) 966-1500

Mailing address: 5165 MacArthur Blvd, NW, Washington, DC 20016

National Filter Media Corp.

Description:	A manufacturer of filter products, pioneering in the manufacture and supply of cotton and other fabrics for filtration purposes, and in the development and application of modern synthetics, such as Acrylic, Polyester, Polypropylene, Nomex, Nylon, Glass, Teflon and Polyethylene. As soon as a synthetic fiber is born, the company tests it for weavability and strength, and for corrosion and temperature resistance when these are unknown factors. New fibers that pass specifications join the line of specialty fabrics rolling off the looms into chemical, food, mining, ceramics, plating and other industries, wherever filtration is employed.
Internet via-web:	http://www.xmission.com/~nfm/
E-mail:	nfm@xmission.com
Phone:	(801) 363-6736; (800) 777-4248
Fax:	(801) 531-1293
Mailing address:	691 North 400 West, Salt Lake City, UT 84103

National Institute of Environmental Health (NIEHS)

Description:	NIEHS is an agency within the National Institute of Health (NIH), which, in turn, is a part of the Department of Health and Human Services. Human health and human disease result from three interactive elements: environmental factors, individual susceptibility, and age. The mission of the National Institute of Environmental Health Sciences (NIEHS) is to reduce the burden of human illness and dysfunction from environmental causes by understanding each of these elements and how they interrelate. NIEHS achieves their mission through multi-disciplinary biomedical research programs, prevention and intervention efforts, and communication strategies that encompass training, education, technology transfer, and community outreach.
Internet via-web:	http://www.niehs.nih.gov
Internet via-gopher:	gopher://gopher.niehs.nih.gov
E-mail:	www@niehs.nih.gov
Mailing address:	Research Triangle Park, NC

National Institutes of Health (NIH)

Description:	NIH uncovers new knowledge that will lead to better health for everyone. NIH works toward that mission by conducting research in their own laboratories; supporting the research of nonfederal scientists in universities, medical schools, hospitals, and research institutions throughout the country and abroad; helping in the training of research investigators; and fostering communication of biomedical information. The NIH is one of eight health agencies of the Public Health Service which, in turn, is part of the U.S. Department of Health and Human Services.
Internet via-web:	http://www.nih.gov
Internet via-gopher:	gopher://gopher.nih.gov
E-mail:	not found
Phone:	not found
Fax:	not found
Mailing address:	Bethesda, MD 20892

National Institute of Occupational Safety and Health (NIOSH)

Description:	NIOSH is a federal agency established by the Occupational Safety and Health Act of 1970. NIOSH is part of the Centers for Disease Control and Prevention (CDC) and is responsible for conducting research and making recommendations for the prevention of work-related illness and injuries.
Internet via-web:	http://www.cdc.gov/niosh
E-mail:	pubstaff@niosdt1.em.cdc.gov

National Institute of Standards and Technology (NIST)

Description:	NIST is a government agency dedicated to a wide range of advanced technology research and development.
Internet via-web:	http://www.nist.gov
E-mail:	webmaster@nist.gov
Mailing address:	Gaithersburg, MD 20899-0001

National Key Centre for Design

Description: A research institute that is maintained by the Royal Melbourne Institute of Technology in Australia. Focuses on the design issues related to pollution prevention.

Internet via-web: http://daedalus.edc.rmit.edu.au/

E-mail: cfd@rmit.edu.au

Phone: 613 9660-2362

Mailing address: Royal Melbourne Institute of Technology, PO Box 2476V, Melbourne Victoria 3001, Australia

National Library of Medicine (NLM)

Description: NLM, located on the southeast corner of the campus of the U.S. National Institutes of Health (NIH) in Bethesda, Maryland, is the world's largest library dealing with a single scientific/professional topic. It cares for over 4.5 million holdings (including books, journal, reports, manuscripts, and audio-visual items). The NLM offers extensive on-line information services (dealing with clinical care, toxicology, and environmental health, and basic biomedical research); has several active research and development components (including an extramural grants program); houses an extensive History of Medicine collection, and provides several programs designed to improve the nation's medical library system.

Internet via-www: http://www.nlm.nih.gov

Internet via-gopher: gopher://gopher.nlm.nih.gov

E-mail: mms@nlm.nih.gov

Phone: (301) 496-0822

Mailing address: Bethesda, MD 20894

National Oceanic and Atmospheric Administration (NOAA)

Description:	Environmental Information Services is a government agency dedicated to weather research and weather data collection.
Internet via-web:	http://www.esdim.noaa.gov
Internet via-gopher:	gopher://gopher.noaa.gov
Internet via-gopher:	gopher://gopher.esdim.noaa.gov
E-mail:	help@esdim.noaa.gov
Phone:	(301) 713-0575
Mailing address:	1315 East West Highway, Room 15400, Silver Spring, MD 20910

National Pollution Prevention Center for Higher Education

Description:	A research institute that is maintained by the University of Michigan. Focuses on the development of P2 educational materials (compendia) for university instructional faculty. These materials help faculty incorporate the principles of P2 into existing or new courses. They contain resources for professors as well as assignments for students.
Internet via-web:	http://www.snre.umich.edu/nppc/
E-mail:	nppc@umich.edu
Phone:	(313) 764-1412
Mailing address:	430 E. University Ann Arbor, MI 48109-1115

National Science Foundation (NSF)

Description:	NSF was established under the National Science Foundation Act of 1950 (Public Law 81-507). The Act authorizes and directs NSF to initiate and support (1) basic scientific research and research fundamental to the engineering process, (2) programs to strengthen scientific and engineering research potential, (3) science and engineering education programs at all levels and in all the various fields of science and engineering, and (4) programs that provide a source of information for policy formulation.
Internet via-web:	http://www.nsf.gov
Internet via-gopher:	stis.nsf.gov
E-mail:	stissserv@nsf.gov
Phone:	(703) 306-1234
Mailing address:	4201 Wilson Blvd, Room P15, Arlington, VA 22230

National Technical Information Service (NTIS)

Description:	NTIS is an indispensable resource for government-sponsored information (U.S. and worldwide). NTIS, an agency within the Technology Administration of the Department of Commerce, serves as the nation's largest central resource for scientific, technical, engineering, and business-related information. NTIS provides you access to more than 2.7 million titles--reports describing research conducted or sponsored by federal agencies and their contractors, statistical and business information, audiovisual products, computer software and electronic databases developed by federal agencies, and technical reports prepared by international research organizations. Approximately 85,000 new titles are added and indexed into the collection annually.
Internet via-web:	http:/www.fedworld.gov/ntis/ntishome.html
E-mail:	orders@ntis.fedworld.gov
Phone:	(800) 553-NTIS; (703) 487-4650
Mailing address:	5285 Port Royal Road, Springfield, VA 22161

National Technology Transfer Center (NTTC)

Description: A uniform resource locator (URL) directory that hosts a variety of servers linking U.S. companies with federal technologies. Those technologies can be converted into practical, commercially-relevant applications. The center's free Gateway Service provides callers with direct contacts in the federal laboratory system. The Gateway Service is available 8:30am - 8pm EST weekdays by calling 800-678-6882.

Internet via-web: http://iridium.nttc.edu

E-mail: webmaster@nttc.edu

Phone: (800) 678-6882

Mailing address: 316 Washington Ave., Wheeling, WV 26003

Nest Group, Inc.

Description: A provider of various chromatography-related products. The company provides kits for DNA, FPLC columns, HPLC columns, etc.

Internet via-web: http://world.std.com/~nestgrp/

E-mail: nestgrp@world.std.com

Phone: (508) 481-6223; (800) 347-6378

Fax: (508) 485-5736

Mailing address: 45 Valley Road, Southborough, MA 01772

Neste Resins North America

Description: A developer and manufacturer of adhesives and resins.

Internet via-web: http://www.neste-resins.com

E-mail: jstone@neste-resins.com

Phone: (541) 687-8840

Fax: (541) 683-1891

Mailing address: 1600 Valley River Drive, Suite 390, Eugene, Oregon

Nevada Technical Associates, Inc.

Description: A provider of training, consulting, and services in radiation protection, radiochemistry, and related areas. The contents of the course include radiation safety officer, gamma spectroscopy, quality assurance in radiochemistry, etc.

Internet via-web: http://www.ntanet.net/

E-mail: nta@ntanet.net

Phone: (702) 564-2798

Mailing address: PO Box 90748, Henderson, NV 89009

Nicolet Instruments

Description: A provider of Fourier transform infrared (FT-IR) and Fourier transform Raman (FT-Raman) spectroscopy products. Also provides information on events, seminars, and trade shows.

Internet via-web: http://www.nicolet.com/

E-mail: nicinfo@nicolet.com

Phone: (800) 232-1472

Fax: (608) 273-5046

Mailing address: 5225 Verona Road, Madison, WI 53711

Norquay Technology Inc.

Description: A manufacturer of various chemicals. Specific chemicals of interest can be found by searching through the Internet homepage, WWW Chemicals, at the Internet address, "http://www3.ios.com:80/~ilyak."

Internet via-web: http://www3.ios.com:80/~ilyak/ind004.html

Phone: (610) 874-4330

Fax: (610) 874-3575

Mailing address: PO Box 468, Chester, PA 19016

North American Catalysis Society

Description: A professional organization for the development of the science of catalysis and related scientific disciplines. Their mission is also to provide educational services to members and other interested individuals; to organize and participate in professional meetings of scientists; to report, discuss, and exchange information and viewpoints in the field of catalysis; to serve as a central exchange for the several catalysis clubs concerning information on their activities; and to provide liaison with foreign catalysis societies, with the International Congress on Catalysis, and with other scientific organizations and individuals, no pecuniary gain or profit to members, incidental or otherwise, being contemplated.

Internet via-web: http://www.dupont.com/nacs/

NSF International

Description: An independent, not-for-profit organization that develops standards, then tests and evaluates products and materials to determine compliance with those standards. Products that meet all requirements of the appropriate standards are then certified by NSF and allowed to carry the NSF Mark on the product, packaging, and/or literature. Since its inception in 1944, NSF has dedicated its work to preserving public health and promoting environmental quality. Today, millions of products around the world bear the NSF Mark.

Internet via-web: http://www.nsf.com/

Phone: (800) NSF-MARK

Mailing address: NSF International, 3475 Plymouth Rd. PO Box 130140, Ann Arbor, MI 48113-0140

Nucleic Acid Database (NDB) Archive

Description: This is a gopher menu. It links to nucleic acid data base newsletters, information ring diagrams and torsion reports, NDB coordinate files, NDB reports in both Ascii format and PostScript format, PDB coordinate files, etc.

Internet via-web: gopher://ndbserver.rutgers.edu:70/11/etc/ndb_link_files

Nutting Environmental of Florida

Description: A consulting firm that provides environmental engineering, consulting, and testing services.

Internet via-web: http://www.gate.net/~nutting/

E-mail: nutting@gate.net

Phone: (407) 737-9700

Fax: (407) 735-4156

Mailing address: 1310 Neptune Drive, Boynton Beach, Florida 33426

O2 Global Network

Description: A uniform resource locator (URL) directory that hosts a variety of server lists related to O_2 issues.

Internet via-web: http://www.wmin.ac.uk/media/O2/O2_Home.html

Occupational Safety and Health Administration (OSHA)

Description: OSHA's mission is to save lives, prevent injuries, and protect the health of America's workers. To accomplish this, federal and state governments must work in partnership with the more than 100 million working men and women and their six and a half million employers who are covered by the Occupational Safety and Health Act of 1970.

Internet via-web: http://www.osha.gov

Office of Industrial Productivity and Energy Assessment

Description: A uniform resource locator (URL) directory that hosts a variety of server lists related to research involving both waste minimization and pollution prevention in industry.

Internet via-web: http://oipea-www.rutgers.edu/html_docs/waste&p2.html

Office of Pollution Prevention and Compliance Assistance

Description: This is an office of the Pennsylvania Department of Environmental Protection. It is looking to form partnerships with Pennsylvania companies and organizations committed to or interested in implementing the ISO 14000 environmental standard.

Internet via-web: http://www.dep.state.pa.us/dep/depurate/pollprev/iso14000/isopart.htm

E-mail: barkanic.robert@a1.dep.state.ps.us

Mailing address: PO Box 2063, Harrisburg, PA 17105-2063

Ohio Valley Plastics Partnership

Description: A plastics engineering company that provides services on manufacturing operations, product design and engineering, ISO 9000 audits, quality systems, materials management, facilities, environmental problems, personnel and training needs, site location and business expansion, and loan and financing information.

Internet via-web: http://www.polysort.com/ohiovaly

Phone: (614) 354-7795

Fax: (614) 353-6353

Old Line Plastics

Description: A manufacturer of plastic products and components for the automotive and consumer products industries.

Internet via-web: http://www.charm.net/~olp/

E-mail: admin@olp1.mhs.compuserve.com

Phone: (410) 879-6010

Fax: (410) 893-0372

Mailing address: 1515 Melrose Lane, Forest Hill, MD 21050-0295

OM Group, Inc. (OMG)

Description:	A producer of metal-based specialty products including metal inorganics, metal powders, metal carboxylates and additives, plastics additives, etc.
Internet via-web:	http://www.usa.net/~omg/
Phone:	(216) 781-8383; (800) 321-9696
Fax:	(216) 781-1502
Mailing address:	3800 Terminal Tower, Cleveland, Ohio 44113

Online Databases, Libraries, and Facilities

DescriptionL	A uniform resource locator (URL) directory that hosts a variety of servers for chemists. It is maintained by the Department of Chemistry, University of California, Davis.
E-mail:	www@chem.ucdavis.edu
Mailing address:	Department of Chemistry, University of California, Davis

Oxford Molecular Group

DescriptionL	A developer and marketer of computer-aided chemistry and bioinformatics software. The company has four major programs: computer-aided molecular design software, bioinformatics tools and related services, computer-aide chemistry software, integrated drug design services.
Internet via-web:	http://www.oxmol.co.uk/
E-mail:	webmaster@oxmol.co.uk

Palm International

Description:	A supplier of various products including chemicals, equipment, metals, instruments, etc.
Internet via-web:	http://www.palminc.com/
E-mail:	palm@metal-finishing.com
Phone:	(615) 793-1990
Fax:	(615) 793-1995
Mailing address:	1289 Bridgestone Pkwy, LaVergne, TN 37086

Papros Inc.

Description: A software company that engages in the application of robust computerized systems for the chemical and environmental planning industries.

Internet via-web: http://www.papros.com

Phone: (408) 432-0126

Fax: (408) 433-5950

Mailing address: 2355 Oakland road, Suite 14, San Jose, CA 95131

Park Equipment Company

Description: A manufacturer of wastewater treatment equipment. Their products include bar screen structures, macerator / grinder pumps, acid neutralization systems, grease interceptors, oil-water separators, sand-oil interceptors, lint interceptors, stormwater interceptors, meter vault assemblies, fire service assemblies, backflow prevention assemblies, surge tank / control valves, dual containment, reclaim tanks, pump lift stations, catchbasins / inlets, manholes / wet wells, sewage treatment systems, bar screen structures, macerator / grinder pumps, and acid neutralization systems.

Internet via-web: http://rampages.onramp.net/~parkco/

Phone: (800) 256-8041

Park Scientific Instruments

Description: A plastics engineering company that specializes in scanning probe microscopy (SPM) technology.

Internet via-web: http://www.park.com/

Phone: (408) 747-1600

Fax: (408) 747-1601

Mailing address: 1171 Borrengas Ave, Sunnyvale, CA 94089

Parr-Green

Description: A plastics engineering company that offers quality tooling for plastics injection, aluminum and zinc die casting dies.

Internet via-web: http://www.polysort.com/parrgreen

Phone: (216) 499-4913

Fax: (216) 499-4598

Mailing address: 9107 Pleasantwood Avenue, NW North Canton, OH 44720

Performance Plastics

Description: A plastics engineering company that offers services on complete plastics engineering services, design consultation, material selection, tool design and construction, and precision component molding.

Internet via-web: http://www.polysort.com/PPI

Phone: (513) 321-8404

Fax: (513) 321-0288

Mailing address: 4435 Brownway Avenue, Cincinnati, OH 45209

Perkin-Elmer Corporation

Description: A supplier of analytical, bioresearch, environmental, and process analytical systems for research, analysis, quality assurance, and related applications. These instrument systems have wide application in the company's key business markets including genetic analysis, environmental monitoring, process analysis, and pharmaceutical quality assurance.

Internet via-web: http://www.perkin-elmer.com/

E-mail: webmaster@perkin-elmer.com

Phone: (203) 762-1000; (800) 762-4000

Fax: (203) 762-6000

Mailing address: 761 Main Ave., Norwalk, CT 06859-0001

Pfizer International

Description: A research-based, global health care company. The Company has four business segments: health care, animal care, consumer health care, and food science.

Internet via-web: http://www.cyber.nl/pfizer/

Phone: 010-406.42.00

Fax: 010-406.42.99

Mailing address: Roer 266, Capelle a/d IJssel, The Netherlands

Pharm-Eco Laboratories, Inc

Description: A drug synthesis and chemical services company that performs a variety of laboratory, process scale-up, and manufacturing tasks.

Internet via-web: http://www.biospace.com/pharmeco

E-mail: main@pharmeco.com

Phone: (617) 861-9303

Fax: (617) 861-9386

Mailing address: 128 Spring Street, Lexington, MA 02173

Phoenix Polymers, Inc.

Description: A compounder of engineering thermoplastic alloys. Their products include PD-8100, which is an alloy of polycarbonates and PET (polyethylene terethalate), SD-8192F, which is an alloy of polycarbonates, and PET with a proprietary modified acrylic, etc.

Internet via-web: http://www.iii.net/biz/phoenix.html

E-mail: ghkosk@phoenik.iii.net

Phone: (508) 345-6200

Mailing address: 225 Industrial Road, Fitchburg, MA 01420

Pilot Chemical Company

Description: A manufacturer of specialty surfactants for household, industrial, and institutional and personal care.

Internet via-web: http://www.pilotchemical.com/welcome.html

E-mail: nburns@pilotchemical.com

Phone: (800) 707-4568, (908) 576-1900

Fax: (908) 530-0844

PIXE Analytical Laboratories, Inc.

Description: An analytical service laboratory specializing in proton-induced X-ray emission spectroscopy (PIXE). This technique offers simultaneous, nondestructive, multi-element analysis in a variety of sample matrices for the elements from sodium through uranium.

Internet via-web: http://www.supernet.net/~pixe/pixe.html

Phone: (904) 576-3900; (800) 700-PIXE

Fax: (904) 576-9076

Mailing address: 1380 Blountstown Highway, Tallassee, FL 32304

Plasti Dip Plastic Spray

Description: A plastics engineering company that offers flexible rubber costing on hundreds of applications to prevent rust and corrosion.

Internet via-web: http://www.cais.com/cytex/pdi/pdi.html

Phone: (800) 969-5432; (612) 785-2156

Fax: (612) 785-2058

Mailing address: 3760 Flowerfield Rd., PO Box 130, Circle Pines, MN 55014

Plastic Bag Association

Description: A professional organization concerned with a broad range of plastic bag issues.

Internet via-web: http://penbiz.com/stellar/green/

Plastic Express

Description: A plastics engineering company that offers shipping of plastic materials.

Internet via-web: http://www.polysort.com/px

Phone: (909) 947-6460

Fax: (909) 947-8164

Mailing address: 2301 East Francis St. Ontario, CA 91761

Plastic Technology Center

Description: A designer, developer, and producer of engineered plastic parts and assemblies. Their products include materials engineering, tooling and processing, software aids, and finite element analysis.

Internet via-web: http://www.lexmark.com/ptc/ptc.html

E-mail: spanoudi@lexmark.com

Phone: (800) 232-7909, (606) 232-7900

Fax: (606) 232-2103

Mailing address: 740 New Circle Road, NW, Lexington, KY 40511

The Plastics Group

Description: A plastics engineering company that offers custom injection molding, product development, testing, and tooling.

Internet via-web: http://www.theplastics.com/tpg

Phone: (770) 513-1769

Fax: (770) 962-4263

Mailing address: 1800 MacLeod Drive, Lawrenceville, GA 30243

Plastics Hotline

Description: A weekly publication of plastics hotline information. The information includes new and used processing equipment, materials, and business opportunities for sale, etc.

Internet via-web: http://www.polysort.com/plasthot

Phone: (800) 247-2000

Fax: (515) 955-6108

Mailing address: 1003 Central Ave., Fort Dodge, IA 50501

Plastics Network

Description: A plastics engineering company that provides a data base center for information, communications, and commerce for the plastics industry.

Internet via-web: http://www.plasticsnet.com/index.html

Polaris Plastic Sales

Description: A plastics engineering company that offers sales of plastic materials.

Internet via-web: http://www.polysort.com/polaris

Phone: (800) 722-5238; (614) 848-4476

Fax: (614) 848-4830

Mailing address: 709 Ozem Gardner Way, Westerville, OH 43081

Pollution Prevention and Waste Minimization

Description: This is a compilation of pollution prevention initiatives by auto companies under the Auto Industry Pollution Prevention Project (Auto Project). The Auto Project is a partnership between the Michigan Department of Environmental Quality and Chrysler, Ford, and General Motors to focus pollution prevention efforts on persistent toxic substances that adversely affect the Great Lakes basin.

Internet via-web: http://www.sme.org/apaa/pollut.html

Pollution Prevention Program Database

Description: This database is maintained by the Pollution Prevention Research Center (PPRC) in Seattle, WA. PPRC supports the development of the database project and acts as a resource center for pollution prevention.

Internet via-gopher: gopher://gopher.pnl.gov:2070/1/.pprc

E-mail: d_leviten@ccmail.pnl.gov

Phone: (206) 223-1151

PolyLinks

Description: An information hub of plastics and polymers. The information hub contains plastics polymers index resource, plastics DotCom magazine, plastics jobbank, etc.

Internet via-web: http://www.polymers.com/

Phone: (508) 345-6200

Mailing address: PO Box 1035, Leominster, MA 01453

PolyNet

Description: For people working in the polycrystalline semiconductor and insulator areas to get together, sharing information and know-how in one of the most challenging fields of Materials Science. Professionals working in the most diverse areas, from crystal growth to photovoltaics, from device physics to spectroscopy, have had a chance to join their ideas and feelings behind the specialty of each field.

Internet via-web: http://www.cilea.it/polynet/

PolySort

Description: A uniform resource locator (URL) directory that hosts a variety of server lists related to polymer industry on the Internet.

Internet via-web: http://www.polysort.com

Powder Page

Description: A uniform resource locator (URL) directory that hosts a variety of servers who are interested in fundamental research of powder. Information provided includes papers, home pages, and addresses.

Internet via-web: http://www.granular.com/

Pressure Chemical Co.

Description: A manufacturer of various chemicals. Specific chemicals of interest can be found by searching through the Internet homepage, WWW Chemicals, at the Internet address, "http://www3.ios.com:80/~ilyak."

Internet via-web: http://www3.ios.com:80/~ilyak/ind025.html

Phone: (412) 682-5882

Fax: (412) 682-5864

Mailing address: 3419 Smallman Street, Pittsburg, PA 15201

Pro-Mold

Description: A manufacturer of custom-designed injection molds and extrusion dies for the plastics and rubber industries.

Internet via-web: http://www.polysort.com/promold

Phone: (216) 920-1560

Mailing address: 1727 Front Street, Cuyahoga Falls, OH 44221

Progressive Products, Inc.

Description: A distributor of Thin Film Technologies (TFT). Their products are solvent-free.

Internet via-web: http://www.tenagra.com/progress/

E-mail: p.oman@ix.netcom.com

Phone: (713) 997-9872

Fax: (713) 997-9895

Mailing address: 4607 Linden Pl., Pearland, Texas 77584

Publishers

Description: A uniform resource locator (URL) directory that hosts a variety of servers in the area of publication activities. The servers provided include Addison Wesley, American Association of University Presses, Blackwell Science, Cambridge University Press, etc.

Internet via-web: http://www.comlab.ox.ac.uk/archive/publishers.html

E-mail: Jonathan.Bowen@comlab.ox.ac.uk

QMR Plastics

Description: A division of Quadion Corporation. It is a world class plastics injection molder.

Internet via-web: http://www.spacestar.com/users/barqmr/

Mailing address: River Falls, WI

Quadrax Corporation

Description: A plastics engineering company that offers thermoplastic composites.

Internet via-web: http://www.growth.com/MENU/QDRX/QDRXhome.html

Phone: (401) 683-6600

Fax: (401) 683-5630

Mailing address: 300 High Point Avenue, Portsmouth, RI 02871

Quality Chemicals Inc.

Description: A provider of custom manufacturing of agricultural, pharmaceutical, polymer, and photoactive chemicals.

Internet via-web: http://www.firstmiss.com/qci/qci_home.html

Phone: (601) 949-0233

Fax: (601) 949-0264

Mailing address: PO Box 1249, Jackson, MS 39215

Questel/Orbit

Description: A French company specializing in patent, trademark, scientific, chemical, business, and news information.

Internet via-web: http://www.questel.orbit.com/patents/oqovw.html

Phone: (800) 456-7248; (703) 442-0900

Fax: (703) 893-4632

Mailing address: France Telecom Group, 8000 Westpark Drive, McLean, VA 22102

Quimica Carnot, S.A.

Description: A manufacturer of raw materials, pharmaceutical products, and premixes to be used in manufacturing several medicines for human or veterinary use. The company is located in Mexico.

Internet via-web: http://www.spin.com.mx/grupocs/gcs-qcab.html

Radiation Research Journal

Description: An official journal of the Radiation Research Society. It comprises a comprehensive source of information for researchers in the radiation sciences.

Internet via-web: http://www.whitlock.com/kcj/science/radres/default.htm

E-mail: radres@aol.com

Mailing address: 2021 Spring Road, Suite 600, Oak Brook, IL 60521

Recycler's World

Description: A provider of an information exchange for those who wish to BUY/SELL/TRADE recyclable commodities, used materials, and collectible items.

Internet via-web: http://www.recycle.net/recycle/index.html

Mailing address: PO Box 24017, Guelph, Ontario, Canada, N1E 6V8

Recycling WWW Servers

Description: A uniform resource locator (URL) directory that hosts a variety of server lists related to waste recycling. Maintained by the University of Windsor.

Internet via-web: http://ie.uwindsor.ca/ecdm_info/recycle.html

Reid Engineering

Description: A consulting firm specializing in the design and operation of water and wastewater treatment systems for industrial, municipal, and private development projects.

Internet via-web: http://users.aol.com/reidengnr/reid.html

E-mail: ReidEngnr@aol.com

Phone: (540) 371-8500

Fax: (540) 371-8576

Mailing address: 1210 Princess Anne Street, Fredericksburg, VA 22401

Reilly Industries, Inc.

Description: A manufacturer of various chemicals. Specific chemicals of interest can be found by searching through the Internet homepage, WWW Chemicals, at the Internet address, "http://www3.ios.com:80/~ilyak."

Internet via-web: http://www3.ios.com:80/~ilyak/ind005.html

Phone: (317) 247-8141

Fax: (317) 248-6413

Mailing address: 1500 South Tibbs Ave., PO Box 42912, Indianapolis, IN 46242

Remco Engineering

Description: A supplier of both influent and effluent water treatment equipment and services to the electronics and allied plating industries.

Internet via-web: http://www.remco.com./~remcobob/home.htm

Phone: (805) 640-0086

Replas

Description: A plastics engineering company that is a midwestern-based custom compounder specializing in commodity and engineered thermoplastics.

Internet via-web: http://www.replas.com/

Phone: (812) 421-3600

Mailing address: 421 N Main St., Evansville, IN 47711

Rescan International Inc.

Description: A consulting firm that provides solutions to a broad spectrum of environmental problems associated with resource development.

Internet via-web: http://www.rescan.com/res/

E-mail: capelletier@rescan.com

Phone: (206) 624-9188

Fax: (206) 382-9648

Mailing address: Suite 3200, 1001 Fourth Avenue Plaza, Seattle, WA 98154

Research Organics

Description: A chemical manufacturer of biochemicals for diagnostics, biochemical research, and biotechnology. The company is producing more than 2,500 commodity and specialty biochemicals on a scale from milligrams to metric tons.

Internet via-web: http://resorg.com/

E-mail: info@resorg.com

Phone: (216) 321-0570; (800) 321-0570

Fax: (216) 883-1576

Mailing address: 4353 East Street, Cleveland, OH 44125

Reuther Mold & Machine

Description: A plastics engineering company that offers complete design of molds.

Internet via-web: http://www.polysort.com/reuther

Phone: (216) 923-5266

Fax: (216) 923-9930

Mailing address: 1225 Munroe Falls Avenue, Cuyahoga Falls, OH 44222-0148

Reverse Logistics (Take-Back Technology)

Description: A research institute that is maintained by the Eindhoven University of Technology. Focuses on the process of continuously taking back products and/or packaging materials to avoid further waste disposal in landfills or high energy consumption through the incineration process.

Internet via-web: http://www.tue.nl/ivo/lbs/main/index.htm

E-mail: bdaatn@urc.tue.nl

Mailing address: Eindhoven University of Technology

Rhone-Poulenc Surfactants and Specialties

Description: A developer of surfactants, cleaning compounds, and specialty chemicals.

Internet via-web: http://www.rpsurfactants.com

E-mail: info@rpsurfactants

Rhyme Industries

Description: A consulting firm that markets a special type of container that is biodegradable and is made from potato and wheat derivatives.

Internet via-web: http://www.envirolink.org/products/rhyme

Phone: (800) 561-8656

Mailing address: Vancouver, B.C.

Ribbon-Jet Tek

Description: A recycling company that provides services on how to recycle printer ribbons, ink jet cartridges, and some Laser Jet toner cartridges.

Internet via-web: http://usa.net/ca/casual.html

E-mail: stanford@rmii.com

Phone: (719) 636-1155

Ridout Plastics - Award Winning Plastics Company

Description:	A plastics engineering company that offers manufacturing and materials for the plastics industry.
Internet via-web:	http://www.sddt.com/~plastics/
E-mail:	PlasticWiz@aol.com
Phone:	(619) 560-1551
Fax:	(619) 560-1941
Mailing address:	5535 Ruffin Road, San Diego, CA 92123

Ritrama Group

Description:	A plastics engineering company that offers the manufacture and sale of specialty self-adhesive materials. It has been certified to meet ISO 9001 standards.
Internet via-web:	http://galactica.galactica.it/ritrama/
Phone:	039 83-9215
Fax:	039-93-4718
Mailing address:	20052 Monza, Milano, Italy

RJF International Corporation

Description:	A manufacturer of engineered polymer products for select specialty markets. For example, wall coverings in an exclusive luxury hotel, protective linings in a power plant flue gas scrubber, backlit awnings on a popular restaurant, and flooring in the combat information center on board a high-tech Navy ship.
Internet via-web:	http://www.polysort.com/RJF
Phone:	(216) 668-7600
Fax:	(216) 668-7705
Mailing address:	3875 Embassy Parkway, Fairlawn, OH 44333

Ro-Mai Industries

Description: A plastics engineering company that offers state-of-the-art CAD/CAM equipment blended with engineering expertise to facilitate highly complex designs to precise specifications. Products include standard and nonstandard items such as tilt-latches and limit stops.

Internet via-web: http://www.polysort.com/RMI

Phone: (800) 321-3543; (216) 425-9090

Fax: (216) 425-7899

Mailing address: 1900 Enterprise Parkway, Twinsburg, OH 44087

Robert Eller Associates

Description: A plastics engineering company that provides services to plastics and rubber industries.

Internet via-web: http://www.polysort.com/REA

E-mail: jvhl11a@prodigy.com

Phone: (216) 665-1139

Fax: (216) 665-1982

Mailing address: 696 Treecrest Drive, Akron, OH 44333

Rohm and Haas Company

Description: A producer of specialty polymers and biologically active compounds. Their products include polymers and resins, monomers, formulation chemicals, AtoHaas, plastics additives, petroleum chemicals, biocides, separation technologies, electronic chemicals, and agricultural chemicals.

Internet via-web: http://www.rohmhaas.com/

E-mail: Webmaster@RohmHaas.Com

Rothberg, Tamburini and Winsor, Inc.

Description: A consulting firm specializing in a multi-disciplinary water and wastewater treatment.

Internet via-web: http://204.131.74.102/RTW/default.htm

Phone: (303) 825-5999

Mailing address: 1600 Stout Street, Suite 1800, Denver, CO 80204-3126

Sachem, Inc.

Description:	A manufacturer of various chemicals. Specific chemicals of interest can be found by searching through the Internet homepage, WWW Chemicals, at the Internet address, "http://www3.ios.com:80/~ilyak."
Internet via-web:	http://www3.ios.com:80/~ilyak/ind022.html
Phone:	(512) 444-3626
Fax:	(512) 445-5066
Mailing address:	821 E. Woodward, Austin, TX 78704

SAF Bulk Chemicals

Description:	The Bulk Chemicals Division of Sigma, Aldrich and Fluka Chemical Companies. The company offers over 50,000 products for biotechnology, pharmaceutical, and diagnostic manufacturing.
Internet via-web:	http://www.sigma.sial.com/sigma/saf1hp.htm
E-mail:	safusa@sial.com
Phone:	(314) 534-4900; (800) 336-9719
Fax:	(314) 652-0000; (800) 368-4661
Mailing address:	SAF USA, 3050 Spruce Street, St. Louis, MO 63103

Sajar Plastics

Description:	A plastics engineering company that offers standard injection molding, structural foam molding, gas-assist injection molding, and painting and assembly.
Internet via-web:	http://www.polysort.com/sajar
Phone:	(216) 632-5203
Fax:	(216) 632-1848
Mailing address:	PO Box 37, 15285 South State Ave., Middlefield, OH 44062

Sampling Systems by PMMI, Inc.

Description: A designer and manufacturer of liquid and gas sampling equipment to help you comply with EPA, OSHA, and NESHAP regulations. Volatile Organic Chemicals(VOCs) can now be sampled with safety while obtaining the most representative sample possible.

Internet via-web: http://mfginfo.mfginfo.com/mfg/sampling/

E-mail: samplers@sat.net

Phone: (409) 647-4421

Fax: (409) 647-5041

Mailing address: Drawer 567 - County Road 375, Old Ocean, TX 77463

SCC Environmental

Description: A consulting firm. The company provides a variety of environmental services and works closely with industry, government, institutions and community groups to increase the awareness of special waste disposal alternatives and to develop responsible attitudes in accordance with the principles of environmental stewardship.

Internet via-web: http://www.compusult.nf.ca/~pantle/

E-mail: pantle@public.compusult.nf.ca

Phone: (902) 461-9131

Fax: (902) 461-0070

Mailing address: Suite 620, 33 Alderney Drive, Dartmouth, Nova Scotia, Canada, B2Y 2N4

Schroinger, Inc.

Description: A provider of software for solving electronic structure problems. It has developed an approach to solving ab initio electronic structure equations, which involves the use of pseudospectral methods.

Internet via-web: http://www.psgvb.com/

E-mail: info@psgvb.com

Phone: (818) 568-9392; (800) 20-PS-GVB

Fax: (818) 568-9778

Mailing address: 80 South Lake Ave., Suite 735, Pasadena, CA 91101

Scientific Instrument Services (SIS)

Description: A supplier of a wide array of products and services for mass spectrometers, gas chromatographs, liquid chromatographs, and other scientific instruments.

Internet via-web: http://www.sisweb.com/

E-mail: sis@sisweb.com

Phone: (908) 788-5550

Fax: (908) 806-6631

Scott Butner's List of Internet Environmental Sources

Description: A uniform resource locator (URL) directory that hosts a variety of server lists related to green engineering and is maintained by the Battelle Seattle Research Center.

Internet via-web: http://ie.uwindsor.ca/ecdm_info/butner.html

E-mail: butner@battelle.org

Phone: (206) 528-3290

Mailing address: Environment and Society Group, Battelle Seattle Research Center, 4000 NE 41st Street, Seattle, WA 98105

SCS Engineers

Description: A consulting firm specializing in the management of solid wastes and hazardous substances.

Internet via-web: http://204.240.184.66/index.html

E-mail: service@scseng.com

SensonCorp Ltd.

Description: This subsidiary of U.S. Tech manufactures corrosion preventative products. These environmentally friendly rust preventative products offer the solution to the EPA's 1995 phaseout of harmful CFC's and nitrites. Recognizing that environmental pollutants play a very significant role in corrosion of electronic and electrical systems, three International standards were established. They allowed environmental corrosivity to be quantified so that equipment manufacturers and industry users could specify the working environment in which their equipment would operate.

Internet via-web: http://www.usxx.com/

E-mail: ustech@haven.uniserve.com

Phone: (512) 376-1049

Fax: (512) 376-1042

Mailing address: PO Box 697, Lockhart, TX 78644

Sewage in Puget Sound

Description: A municipal wastewater resource. It is about a project report on municipal sewage treatment plants that discharge their effluents into Puget Sound and Washington streams and rivers.

Internet via-web: gopher://futureinfo.com/1/menu3/menu5/

Phone: (206) 382-7007

Mailing address: 1326 Fifth Ave., Suite 450, Seattle, WA 98101

Shell Chemical Company

Description: A major oil company and a manufacturer of various chemicals. Their products include additives, base chemicals, polyketones, detergent, elastomers, polybutylene, polyester, resign, solvents.

Internet via-web: http://www.shellus.com/Chemical/Welcome.html

Phone: (713) 241-6390

Showa Chemical CO., LTD

Description: Provides several information servers on various chemicals. The servers include reagents' chemical data base, MADS's homepage, and other related information.

Internet via-web: http://www.st.rim.or.jp/~shw/

E-mail: shw@st.rim.or.ip

Phone: 03-3270-2720

Fax: 02-449-6457

Sigma Chemical Company

Description: A provider of research chemicals and other specialty products. Their product information includes technical presentations and seminars, selected product areas, enzymatic assay procedures, etc.

Internet via-web: http://www.sigma.sial.com/

E-mail: webmaster@sial.com

SimaPro3

Description; A consulting firm. The company provides services on life cycle analysis tools, with downloadable demonstration, for assisting decision making.

Internet via-web: http://www.ivambv.uva.nl/pre/simapro.html

E-mail: webmaster@ivambv.uva.nl

Phone: 31-33-65-28-53

Mailing address: 3811 NH Amersfoort, The Netherlands

Smith's Enterprises

Description: A plastics engineering company that offers molded urethane products.

Internet via-web: http://www.iquest.net/~ariel/ibd/label13.html

Phone: (317) 378-6267

Fax: (317) 378-6317

Mailing address: 1124 Dilts Street, Chestfield, IN 46017

Smithers Scientific Services

Description: A plastics engineering company that offers management consulting and market research, laboratory testing and technical consulting, vehicle testing and performance evaluation, marine research and testing, tire testing and analysis, industry reports.

Internet via-web: http://www.polysort.com/companies/s/smithers/smithome.htm

Phone: (216) 762-7441

Fax: (216) 762-7447

Mailing address: 425 W. Market Street, Akron, OH 44303-2099

Solstice

Description: A research institute that is sponsored by the U.S. Department of Energy. A multimedia encyclopedia on renewable energy and the environment. The Renewable Energy Exhibit is an overview of renewable energy technologies and their applications. Both CD-ROMs are available for purchase on-line along with CREST ultimate frisbees.

Internet via-web: http://solstice.crest.org/

Solutions Software

Description: A publication company. The company publishes the public domain reference materials including The TSCA Chemical Data Inventory on CD-ROM, the complete U.S. Code of Federal Regulations, Chemical and Solvent Safety (with full MSDS detail) on CD-ROM, Chemical Substitutions and Chemical Compatibility Datafiles on CD-ROM, Innovative Environmental Technology / Treatability Studies on CD-ROM, Material Safety Data Sheets Database (160,000 data sheets) on CD-ROM.

Internet via-web: http://www.env-sol.com/

E-mail: solution@iag.net

Phone: (407) 321-7912

Fax: (407) 321-7912

Mailing address: 1795 Turtle Hill Road, Enterprise, FL 32725

Specialty Recycling Services

Description: A plastics engineering company that supplies the rubber and plastics industry with cryogenically ground scrap and raw polymers.

Internet via-web: http://www.polysort.com/SRS

Phone: (216) 883-7673

Mailing address: 3313 East 80th Street, Cleveland, OH 44127

Spectrocell

Description: A manufacturer of equipment and components for use in the analytical science and research industry. The products include absorption cells for spectrophotometry, colorimetry, and fluorimetry.

Internet via-web: http://www.netaxs.com/~gerryms/spctrcll.html

E-mail: gerryms@netaxs.com

Phone: (215) 572-7605; (800) 670-2773

Fax: (215) 885-2792

Mailing address: 143 Montgomery Ave., Oreland, PA 19075

Spirex

Description: A plastics engineering company that develops plasticating components. Their products include plasticating screws, nonreturn valves, software development, custom lab testing, hydraulic oil filtration, and spirex history.

Internet via-web: http://www.polysort.com/spirex

E-mail: spirex1@aol.com

Phone: (216) 726-1168

Fax: (216) 726-9437

Mailing address: 8469 Southern Blvd. PO Box 9130, Youngstown, OH 44512

SRI International

Description: Founded in 1946 in conjunction with Stanford University as the Stanford Research Institute, later became fully independent and were incorporated as a nonprofit organization under U.S. and California laws. Originally founded to provide a center for diversified multi-disciplinary scientific research, SRI serves clients in the United States and throughout the world. Research is SRI's only product. SRI does not manufacture or market any product, nor is it connected with any manufacturing organization. Thus, SRI is able to provide research analyses and system designs unbiased by any proprietary interest in specific areas of technology. About 1,000 research projects are under way at any given time; the current volume of contract research is over $200 million per year.

Internet via-web: http://www.sri.com/

Mailing address: Headquarters and principal laboratories are in Menlo Park, California

STN International

Description: A scientific and technical information network. This network is operated cooperatively by Fachinformationszentrum (FIZ) Karlsruhe, Chemical Abstracts Service (CAS) of the American Chemical Society (ACS), and the Japan Information Center of Science and Technology (JICST). Service centers in Karlsruhe, Columbus and Tokyo are linked by sea cable (TAT8). Thus, the users have access to one worldwide information service with up-to-date databases in science and technology. Describes activities of the laboratory and the people involved.

Internet via-web: http://www.fiz-karlsruhe.de/stn.html

STN International the Scientific and Technical Information Network

Description: The network under PDLCOM was produced by the Plastics Design Library (PDL). It contains test data on the chemical compatibility and the environmental stress crack resistance of plastics. Properties include the effect of about 3300 chemical reagents, radiation, heat, and outdoor environment with or without additional stress on the mechanical properties, geometry, weight, and appearance of over 400 plastics. It also contains a thesaurus capability that provides definitions, hierarchical information, and synonyms for property terms, generic families, and chemical types of plastics and fillers within the file.

Internet via-web: http://www.fiz-karlsruhe.de/pdlcom.html

Phone: (212) 838-2817

Mailing address: William Andrew, Inc., Plastics Design Library, 345 East 54th Street, Suite 5E, New York, NY 10022

Stormceptor,Inc.

Description: A developer of a stormwater quality manhole interceptor that is designed to remove suspended solids and floatables from urban runoff.

Internet via-web: http://www.stormceptor.ca

E-mail: gbryant@stormceptor.ca

Phone: (800) 762-4703 (USA); (800) 565-4801 (Canada)

Struktol Company of America

Description: A plastics engineering company that introduces processing and homogenizing agents to tire compounders.

Internet via-web: http://www.polysort.com/companies/s/struktol/strkhome.htm

Phone: (216) 928-5188

Fax: (216) 928-8726

Mailing address: 201 East Steels Corners Road, PO Box 1649, Stow, OH 44224-0649

Sumitomo Chemical Co.

Description: Established in 1913, one of Japan's leading chemical manufacturers. The company has obtained ISO 9002 certificates for all company's manufacturing homepages. Their products includes basic chemicals, petrochemicals, fine chemicals, agricultural chemicals, and pharmaceuticals.

Internet via-web: http://www.sumitomo-chem.co.jp/

E-mail: web@www.sumitomo-chem.co.jp

Phone: 03-5534-5500

Mailing address: 27-1 Shinkawa 2-chome, Chuo-ku, Tokyo 104 Japan

Sustainable Development Online

Description: A uniform resource locator (URL) directory that hosts a variety of server lists related to sustainable development. Subjects include Introduction to ISO, ISO technical committees, ISO structure, ISO meeting calendar, ISO members worldwide, ISO Catalogue, ISO 9000 News Service, What's new at ISO?, and other web servers providing standards information.

Internet via-web: http://www.iso.ch/

Synthetech

Description: A manufacturer of various chemicals. Specific chemicals of interest can be found by searching through the Internet homepage, WWW Chemicals, at the Internet address, "http://www3.ios.com:80/~ilyak."

Internet via-web: http://www3.ios.com:80/~ilyak/ind011.html

Phone: (503) 967-6575

Fax: (503) 967-9424

Mailing address: 1290 Industrial Way, PO Box 646, Albany, Oregon 97321

Systems Realization Laboratory

Description: A research institute that is maintained by the Georgia Institute of Technology. The focus is on the decision-based design and realization of open and sustainable engineering systems. In this context, the research activities are directed towards the development of computer-based methods and tools for designing, producing, deploying, maintaining, and retiring engineering systems in technically, economically, ecologically, and ethically sound ways.

Internet via-web: http://www.srl.gatech.edu/

E-mail: gt7883d@prism.gatech.edu

Phone: (404) 894-8170

Mailing address: School of Mechanical Engineering, Georgia Institute of Technology, Atlanta, GA 38332

Tech Mold

Description: A plastics engineering company that provides a variety of services to the plastics industry.

Internet via-web: http://www.industrylink.com/techmold

E-mail: techmold@indirect.com

Phone: (602) 968-8691

Fax: (602) 968-7359

Mailing address: 1735 West 10th Street, Tempe, Arizona 85281-5295

The Technology, Business and the Environment Program

Description: A research institute that is maintained by the Massachusetts Institute of Technology (MIT). This program helps companies meet the dual challenges of achieving environmental excellence and business success. The program's mission is to elucidate a new preventive environmental management paradigm, centering on business practices and linking technological change with sound environmental management.

Internet via-web: http://web.mit.edu/afs/athena/org/c/ctpid/www/tbe/index.html

Mailing address: Massachusetts Institute of Technology

Technology and Trade Inc.

Description: A plastics engineering company that provides a variety of services to the plastics industry.

Internet via-web: ://www.techtrade.com

E-mail: info@techtrade.com

Phone: (312) 266-7414

Fax: (312) 943-1634

Mailing address: 1559 North LaSalle Street Suite 1800, Chicago, IL 60610

TeleChem International, Inc.

Description: A trade company that provides import, export, and distribution services of chemicals.

Internet via-web: http://hooked.net/users/telechem

E-mail: telechem@hooked.net

Phone: (408) 977-0160

Fax: (408) 977-0164

Mailing address: 12 South First Street, Suite 817, San Jose, CA 95113-2405

Texaco OnLine

Description: A major oil, gas, and petrochemical company with operations in many foreign countries.

Internet via-web: http://www.texaco.com/

Phone: (800) 283-9785

Thermochemical Calculator (TCC)

Description: A WWW tool to carry out thermochemical calculations for ideal gas mixtures. It makes use of Chemkin, and includes a database of properties for many species of interest for combustion, atmospheric chemistry, or chemical vapor deposition problems.

Internet via-web: not foundhttp://adam.caltech.edu/tcc/

Thomas

Description: A uniform resource locator (URL) directory that hosts a variety of servers in the area of congressional activities. The servers provided include full text of legislation, full text of congressional record, congressional record index, etc.

Internet via-web: http://rs9.loc.gov

E-mail: thomas@loc.gov

Thomas Register of American Manufacturers

Description: A uniform resource locator (URL) directory that hosts a variety of servers related to American manufacturers. The Thomas Register has more than 50,000 separate product and service headings.

Internet via-web: http://www.thomasregister.com/

Phone: (800) 222-7900 ext 200

Mailing address: 5 Penn Plaza, NY, NY 10001

Toagosei Co., Ltd.

Description: A Japanese company that deals with a broad range of fields, from material-based products (including caustic soda, vinyl chloride and acrylic esters) to adhesives and specialty chemical products such as acrylic polymers. It has also diversified into bioscience and electronic materials.

Internet via-web: http://www.dir.co.jp/cib/4045/welcome.html

E-mail: cib@daiwa.co.jp

Phone: 03-3597-7215

Mailing address: 14-1 Nishi Shimbashi 1, Minato-ku, Tokyo 105, Japan

Transportation Resources

Description: A uniform resource locator (URL) directory that hosts a variety of servers who are interested in transportation and its related activities. This server is maintained by the Princeton University on the subjects related to all transportations.

Internet via-web: http://dragon.princeton.edu:80/~dhb/

E-mail: dhb@dragon.princeton.edu

TreeEco

Description: A supplier of nontoxic and bio-degradable cleaning products and paper goods. These materials are recycled with high PCW (Post-Consumer Waste), and thus completing the recycling loop.

Internet via-web: http://www.envirolink.org/treeco/

Trends in Analytical Chemistry (TRAC)

Description: A journal publication that provides overviews of new developments in analytical chemistry, and aims to help analytical chemists and other users of analytical techniques explore and orient themselves in fields outside of their particular specialization(s).

Internet via-web: http://www.elsevier.nl:80/catalogue/SA2/205/06040/06040/502695/502695.html

E-mail: usinfo-f@elsevier.com

Phone: (212) 633-3750

Fax: (212) 633-3764

Mailing address: Elsevier Science, PO Box 945, Madison Square Station, NY, NY 10159-0945

Tres English's Sustainable Development Pages

Description: A uniform resource locator (URL) directory that hosts a variety of server lists related to sustainable development and environment.

Internet via-web: http://www.azstarnet.com/~tres/

Tripos, Inc.

Description: A supplier of scientific software and services to facilitate the discovery of new therapeutic and bioactive compounds in the pharmaceutical, biotechnological, chemical, and agrochemical industries.

Internet via-web: http://www.webcom.com/~tripos2/

E-mail: webmaster

Phone: (800) 323-2960

Fax: (314) 647-9241

Mailing address: based in St. Louis, Missouri

Trojan Technologies Inc.

Description: A Canadian manufacturing company specializing in ultraviolet (UV) light applications for water and wastewater disinfection.

Internet via-web: http://www.trojanuv.com/

E-mail: 1schneid@TrojanUV.com

Phone: (519) 457-3400

Fax: (519) 457-3030

Mailing address: 3020 Gore Road, London, Ontario, Canada N5V 4T7

TSD Central (Hazardous Waste Central)

Description: A waste treatment company that provides a one-stop centralized service for the disposal of hazardous waste.

Internet via-web: http://www.tsdx.com/central

U.S. Technology

Description: A manufacturer of plastic abrasive. Information includes U.S. polymedia plastic abrasive, custom engineered for industry; surface cleaning, paint removal, surface finishing and recycling.

Internet via-web: http://www.polysort.com/ustech

Phone: (800) 634-9185; (216) 455-1181

Fax: (216) 445-1191

Mailing address: 220 7th St. S.E., Canton, OH 42702

UCLA CCT

Description: UCLA CCT is the University of California at Los Angeles Center for Clean Technology. The center conducts a wide range of environmental research and related activities including pollution prevention through innovative technologies and process design, combustion and air toxics, water and wastewater treatment, intermedia transport and fate of chemicals in the environment, remediation and restoration, environmental risk reduction and pollution prevention.

Internet via-web: http://cct.seas.ucla.edu/

E-mail: cct@seas.ucla.edu

Mailing address: 7440 Boelter Hall, PO Box 951600, Los Angeles, CA 90095-1600

Ultrasonic Testing OnLine Journal

Description:	An online journal that presents information regarding Ultrasonics for Nondestructive Testing (NDT). The contents include Educational Material, Articles with topical Applications (also for plastics), News Flashes, Buyers' Guide, a Virtual Exhibition, and a Virtual Library.
Internet via-web:	http://www.ultrasonic.de
E-mail:	dieder@echo.teuto.de
Phone:	49/(0)5221/72969
Fax:	49/(0)5221/769731
Mailing address:	Tacheniusweg 8, 32052 Herford, Germany

Umetrics

Description:	A chemical company that specializes in improving and optimizing the quality and performance of industrial products, chemical structure- property relationships, characterization of materials, and relations between characteristics and quality.
Internet via-web:	http://www.it-center.se/umetri/
E-mail:	info@umetrise
Phone:	46-90-154-840
Fax:	46-90-197-685
Mailing address:	Box 7960 S-90719 Umea, Sweden

Unique Tire Recycling

Description:	A waste recycle company that recycles and reduces waste tires into the following basic components for which there is almost unlimited demand: medium crude oil, steel and carbon black. Plastics are transformed into synthetic crude oil. All of these components are then sold to manufacture new products.
Internet via-web:	http://www.kpp.com/unique/unique.htm
Phone:	(800) 333-8084

Universal Environmental Technologies

Description: A developer of the probailer system which can recover gasoline and lighter petroleum products from any configuration of contaminated sites.

Internet via-web: http://www.uet.com/

E-mail: info@uet.com

Phone: (800) 850-8688; (603) 889-8747

Fax: (603) 889-8072

Mailing address: 4 Townsend West, Unit 7, Nashua, NH 03063

Universal Plastics

Description: A distributor, fabricator, and machinist of plastic products. The company is capable of handling many plastics requirements.

Internet via-web: http://www.polysort.com/universal

Phone: (216) 645-6873

Fax: (216) 645-0064

Mailing address: 2587 South Arlington Road, Akron, OH 44319-2086

Utah Water Research Laboratory (UWRL)

Description: The research mission of the UWRL is to enhance and expand scientific understanding and engineering technology for environmental and water resources management with a special focus on problems affecting the State of Utah. Multi-disciplinary faculty work together in environmental engineering, ground water, fluid mechanics and hydraulics, hydrology, natural systems engineering, water education, and water resources management to accomplish this goal.

Internet via-web: http://publish.uwrl.usu.edu/uwrl.gome.html

Mailing address: Utah Water Research Laboratory, USU College of Engineering, Utah State University

Utility, Power, Wastewater Plant's E-mail Directory

Description: A uniform resource locator (URL) directory that hosts a variety of servers related to utility, power, and wastewater plant activities.

Internet via-web: http://www2.best.com/~lidial/plant.htm

E-mail: lidial@ix.netcom.com

Phone: (914) 963-3695

Van Dorn Demag

Description: A manufacturer and distributor of injection molding machines. The company earned its ISO 9001 in 1993.

Internet via-web: http://www.polysort.com/vandorn

Mailing address: Headquartered in Strongsville, Ohio

Varian Analytical Instruments Company

Description: A provider of systems and components for medical, scientific, and industrial markets. Each day more than 80,000 people receive cancer treatments from more than 3,000 Varian medical linear accelerators, treatment simulators, and information management systems in service at hospitals and clinics worldwide.

Internet via-web: http://www.varian.com/

E-mail: webmaster@varian.com

Mailing address: Palo Alto, California

VERONICA

Description: A uniform resource locator (URL) directory that hosts a variety of gopher addresses. Its gopher menu includes find gopher directories by title word(s), search gopherspace by title word(s), etc.

Internet via-gopher: gopher://veronica.scs.unr.edu/11/veronica

Viscona Limited

Description: A producer of stabilized chlorine dioxide together with associated R&D activities. This is a product with uses within the water treatment industry and also as an odor control agent.

Internet via-web: http://www.scotborders.co.uk/horizons/viscona.html

Mailing address: St. Boswells, Melrose, Scotland, UK

Vista Chemical Company

Description: An integrated producer of commodity and specialty chemicals. Products include linear alcohols and derivatives, polyvinyl chloride resins and compounds, polymer blends and alloys, detergent alkylate, high-purity alumina, and other industrial chemicals.

Internet via-web: http://www.vistachem.com/

Phone: (713) 588-3000

Mailing address: 900 Threadneedle, Houston, TX 77079

VPI

Description: A plastics engineering company that provides a variety of services for the plastics industry.

Internet via-web: http://www.dataplusnet.com/vpi/vpi.html

Phone: (414) 458-4664

Fax: (414) 458-1368

Mailing address: 3123 S. 9th Street, Sheboygan, WI 53081

Wastewater Engineering

Description: A uniform resource locator (URL) directory that hosts a variety of servers in the area of wastewater engineering activities. The servers provided include (1) municipal wastewater resources, (2) academic and research institutions, (3) professional and trade organizations, (4) commercial organizations, (5) other applicable resources, (6) an invitation to submit items.

Internet via-web: http://www.halcyon/cleanh2o/ww

E-mail: cleanh2o@halcyon.com

Mailing address: Industrial Wastewater Engineering, Seattle, Washington

Wastewater Information

Description: A uniform resource locator (URL) directory that hosts a variety of servers related to the following areas: 1) wastewater discharge regulations, 2) chemical precipitation, 3) a wastewater treatment implementation approach outline, 4) a wastewater treatment book and periodical list, 5) a cooling tower water treatment thread, 6) a wastewater management plan, 7) an association and organization list, 8) solutions software, 9) environmental glossary search for wastewater; 10) industrial wastewater treatment--a guidebook.

Internet via-web: http://www.halcyon.com/cleanh2o/ww/wwtinfo.html

E-mail: cleanh2o@halcyn.com

Mailing address: Industrial Wastewater Engineering, Seattle, WA

Wastewater Solutions

Description: Waste biotreatment consultants for manufacturing facilities, R&D management, and A/E and legal firms.

Internet via-web: http://pages.prodigy.com/MO/wastewater/company.html

E-mail: BTXM97@prodigy.com

Water & Wastewater Utilities

Description: A uniform resource locator (URL) directory that hosts a variety of servers related to both private and publicly owned utility companies and districts sites.

Internet via-web: http://www.waterweb.com/ww/utility.html

E-mail: water@amug.org

Phone: (602) 948-3555

Fax: (602) 948-1910

Mailing address: 7905 E. Greenway Road, Suite 106, Scottsdale, AZ

Water Resources Management, Inc.

Description: A consulting firm specializing in the use of mathematical and economic analysis techniques to develop management strategies and define operating policies for water resources systems.
Internet via-web: http://www.wrmi.com/pub/wrmi/wrmi.htm
E-mail: wrmi@wrmi.com
Phone: (916) 920-1811
Mailing address: 1851 Heritage Lane, Suite 101, Sacramento, CA 95815

Water - Wastewater Web

Description: A uniform resource locator (URL) directory that hosts a variety of servers related to wastewater treatment. The servers provided include (1) a showcase for water and wastewater equipment manufacturers, (2) a reference for water and wastewater professionals, and (3) a service for cities, government and industry.
Internet via-web: http://www.w-ww.com
E-mail: webmaster@w-ww.com

WaterOnLine

Description: A uniform resource locator (URL) directory that hosts a variety of servers related to water professionals' activities from around the world. Water and wastewater technology information is continually updated.
Internet via-web: http://www.wateronline.com/
Phone: (800) 324-3951

Waters Corporation

Description: A manufacturer of analytical instruments and chromatography chemistries. The company is devoted to High Performance Liquid Chromatography (HPLC) technology.
Internet via-web: http://www.waters.com/
E-mail: info@waters.com
Phone: (508) 478-2000; (800) 252-4752, (508) 478-2000
Fax: (508) 872-1990
Mailing address: 34 Maple Street, Milford, MA 01757

Waterweb

Description: A uniform resource locator (URL) directory that hosts a variety of servers related to the water technology community.

Internet via-web: http://www.waterweb.com

E-mail: mailto:water@amug.org

Phone: (602) 948-3555

Mailing address: 7905 E. Greenway Road, Suite 106, Scottsdale, AZ

Waterworld

Description: A uniform resource locator (URL) directory that hosts a variety of servers in the area of wastewater treatment activities. It contains product and literature releases that have appeared in WaterWorld beginning with the January 1995 issue.

Internet via-web: http://www.waterworld.com/

E-mail: waterwld@pennwell.com

WAU Process Engineering

Description: A research institute that focuses on nitrogen removal from liquid or gas streams by means of artificially immobilized micro-organisms in airlift loop reactors. Maintained by the Wageningen Agricultural University in the Netherlands.

Internet via-web: http://www.spb.wau.nl/prock/rene.html

E-mail: Phone: 31-0317-482-884

Mailing address: PO Box 8129, 6700 EV Wageningen, The Netherlands

Wavefunction, Inc.

Description: A provider of computational chemistry and molecular modeling, both in research and education. Their products include Spartan, which is a computational chemistry software for UNIX workstations, and MacSpartan, which is a software for the Macintosh computer systems.

Internet via-web: http://www.wavefun.com/

E-mail: access_codes@wavefun.com

Phone: (714) 955-2120

Mailing address: 18401 Von Karman Ave., Suite 370, Irvine, CA 92715

Wiltec Research Company, Inc.

Description:	A consulting firm specializing in measuring thermodynamic properties and in thermodynamics consulting.
Internet via-web:	http://www.xmission.com/~ogdenc/wiltec.html
Phone:	(801) 474-6648
Fax:	(801) 374-6674
Mailing address:	488 South 500 West, Provo, UT 84601

WindowChem Software, Inc.

Description:	A provider of window-based chemistry/laboratory software. The company promotes different softwares from different authors in one catalog. Their goal is to help chemists help chemists through Windows.
Internet via-web:	http://www.windowchem.com/
E-mail:	steve@windowchem.com
Phone:	(707) 864-0845; (800) 536-0404
Fax:	(707) 864-2815
Mailing address:	420-F Executive Court North, Fairfield, CA 94585

Witt Plastics

Description:	A plastics engineering company that provides services to thermoforming, printing, die-cut and lighting industries.
Internet via-web:	http://lakeland.tsolv.com/~wittpl/
Phone:	(800) 867-9488; (941) 665-6550
Fax:	(941) 667-0852
Mailing address:	3010-15 Maine Ave., Lakeland, FL 33801

WMX Technologies

Description: A consulting firm that provides comprehensive environmental, waste management and related services. This company is in partnership with Waste Management, Inc., Chemical Waste Management, Inc., Wheelabrator Technologies Inc., Rust International Inc., and Waste Management International PLC.

Internet via-web: http://www.wmx.com/

Phone: (800) 4WMXNEWS

Mailing address: 3003 Butterfield Road, Oak Brook, IL 60521

World Wide Chemnet, Inc.

Description: An information hub of chemical products. The goal is to help companies and people find chemicals, oils, petrochemicals, solvents, gasses, and raw materials.

Internet via-web: http://www.galstar.com/~chemnet/index.html

E-mail: info@netscape.com

Phone: (918) 749-9056

Fax: (918) 747-1444

Mailing address: 2424 E. 21st Street, Ste 450, Tulsa, OK 74114

WWTNET

Description: The Wastewater Treatment Network (WWTNET) is a group of environmental professionals who specialize in the pretreatment and secondary treatment of industrial and municipal wastewaters. WWTNET is a member contribution forum. It was created as a means of establishing a forum of professionals and interested parties that can share opinions, knowledge, problems, and technologies used in the treatment of wastewaters.

Internet via-web: http://users.aol.com/wwtnet/wwtnet.html

E-mail: wwtnet@aol.com

WWW Chemicals

Description: A provider of an online system from which information on a specific chemical (e.g., its manufacturer and MSDS data) can be retrieved electronically. The system combines product catalogs of over twenty chemical companies. Their product listings range from several dozens of specialty chemicals manufactured by H&S Chemical to over 12,000 synthetic reagents and chemical building blocks provided by Lancaster Synthesis. You can search the entire WWW Chemicals catalog, or any of individual companies' catalogs.

Internet via-web: http://www3.ios.com:80/~ilyak

E-mail: ilyak@soho.ios.com

Phone: (714) 993-5232

Fax: (714) 993-1471

Mailing address: 1431 Avenida Alvarado, Placentia, CA 92670

Zefon Manufacturing

Description: A plastics engineering company that manufactures medical, environmental, and plastic products. Their products include (1) Zefon analytical accessories, which serves the environmental industry with air sampling pumps and cassettes and is a distributor of accessories; (2) Zefon manufacturing, which serves a variety of industries with custom assembly, packaging, and injection molding; (3) Zefon medical products, which serves the medical device industry with disposable medical products.

Internet via-web: http://www.zefon.com/

E-mail: zefon@zefon.com

Phone: (800) 282-0073

Fax: (352) 854-8080

Mailing address: 2860 23rd Ave N, St. Petersburg, FL 33713

Zipperling Kessler & Co

Description: An information source of masterbatches, compounds, organic metals, and research. Their products include corrosion protection systems, flame retardance materials, etc.

Internet via-web: http://www.http://www.zipperling.de/

CHAPTER 2

WORLD WIDE WEB RESOURCES BY SUBJECT

Chapter 2

World Wide Web Resources by Subject

Chemical Analytical Instrumentation Manufacturers

Alitea USA	http://www.flowinjection.com/flowinjection
ALToptronic AB	http://www.altptronic.se
ASD	http://pluto.njcc.com/~bpapp/chempoin.html
Beckman Instruments, Inc.	http://www.beckman.com/
Catalytica, Inc.	http://www.catalytica-inc.com/
Gilson, Inc.	http://pubs.acs.org/pin/gilson/gil222p1.html
Hewlett-Packard Analytical	http://www.hp.com/go/analytical
Hitachi Instruments, Inc.	http://www.hii.hitachi.com/
Indigo Instruments	http://ds.internic.net/indigo/index.html
LECO Corporation	http://pubs.acs.org/pin/leco/lec.html
Microanalytics Instrumentation	http://www.mdgc.com/
Micromeritics Instrument Corp.	http://www.micromeritics.com/
Nest Group, Inc.	http://world.std.com/~nestgrp/
Nicolet Instruments	http://www.nicolet.com/
Perkin-Elmer Corporation	http://www.perkin-elmer.com/
Sampling Systems by PMMI, Inc.	http://mfginfo.mfginfo.com/mfg/sampling/
Scientific Instrument Services (SIS)	http://www.sisweb.com/

Chemical Analytical Instrumentation Manufacturers (*continued*)

Spectrocell	http://www.netaxs.com/~gerryms/spctrcll.html
Varian Analytical Instruments	http://www.varian.com/
Waters Corporation	http://www.waters.com/

Chemical Analytical Methods, Services, and Information

Alpha Analytical Labs	http://world.std.com/~alphalab/
Analytical Chemistry and Chemometrics Index	http://www.chemie.fu-berlin.de/chemistry/index/anal/
Analytical Chemistry and Instrumentation	http://www.chem.vt.edu/chem-ed/analytical/ac-methods.html
Analytical Chemistry Basics	http://www.chem.vt.edu/chem-ed/analytical/ac-basics.html
Analytical Chemistry Hypermedia	http://www.chem.vt.edu/chem-ed/analytical/ac-home.html
ATI Orion	http://pubs.acs.org/pin/orion/ori.html
Garden State Laboratories	http://www.planet.net/gsl/
Info-Labview Mailing List	ftp://ftp.natinst.com/README
NSF International	http://www.nsf.com/
PIXE Analytical Laboratories, Inc.	http://www.supernet.net/~pixe/pixe.html
Trends in Analytical Chemistry (TRAC)	http://www.elsevier.nl:80/catalogue/SA2/205/06040/06040/502695/502695.html
Ultrasonic Testing OnLine Journal	http://www.ultrasonic.de

Chemical Analytical Methods, Services, and Information *(continued)*

Wiltec Research Company, Inc.	http://www.xmission.com/~ogdenc/wiltec.html

Chemical Computer Software Analysis

ARSoftware's Online Internet Catalog	http://arsoftware.arclch.com/
Aspen Technology	http://www.aspentec.com/
BioSupplyNet	http://www.biosupplynet.com/bsn/
CambridgeSoft, Corp.	http://www.camsci.com/
ChemInnovation Software	http://www.cheminnovation.com/
DAYLIGHT Chemical Information Systems, Inc.	http://www.daylight.com/
Falcon Software	http://www.falconSoftware.com/falconweb/index.html
Hypercube, Inc.	http://www.hyper.com/
IBM World Wide Web	http://www.ibm.com/
Jandel Scientific	http://www.jandel.com/
MDL Informations Systems, Inc.	http://www.mdli.com/
MicroMath Scientific Software, Inc.	http://www.micromath.com/~mminfo/
Molecular Simulations, Inc.	http://www.msi.com/
Oxford Molecular Group	http://www.oxmol.co.uk/
Schroinger, Inc.	http://www.psgvb.com/
SoftShell Online	http://www.softshell.com/
Tripos, Inc.	http://www.webcom.com/~tripos2/

Chemical Computer Software Analysis (*continued*)

Wavefunction, Inc.	http://www.wavefun.com/
WindowChem Software, Inc.	http://www.windowchem.com/

CCA

Chemical Concepts	http://www.vchgroup.de/cc/

Chemical Information Service

Bio-Online	http://www.bio.com/companies/co-info.toc.html
BUBL WWW Subject Tree - Chemical Engineering	http://www.bubl.bath.ac.uk/BUBL/Chemeng.html
CAD Centre at Strathclyde University	http://www.cad.strath.ac.uk/Home.html
Chemical Abstracts Service	http://info.cas.org/welcome.html
Chemical Engineering Sites all over the World	http://www.ciw.unkarlsruhe.de/siteworl.html
Chemical Engineering Sites all over the World	http://www.ciw.ukarlsruhe.de/germany.html
Chemical Physics Preprint Database	http://www.chem.brown.edu/chem-ph.html
Chemical Week Magazine	http://www.chemweek.com/
Chemistry Index / Chemie-Index (FU Berlin)	http://www.chemie.fuberlin.de/chemistry/index.html
Chemistry on the Internet: The Best of the Web 1995	http://www.ch.ic.ac.uk/infobahn/boc.html
ChemKey Database	http://euch6f.chem.emory.edu/

Chemical Information Service *(continued)*

ChemSOLVE	http://www.eden.com/~chemsolv/
GENBBB (Generic Bulletin Board Builder)	http://www.cs.colorado.edu/homes/mcbryan/public_html/bb/167/summary.html
Guide to Chemical Engineering	http://www.theworld.com/SCIENCE/ENGINEER/CHEMICAL/SUBJECT.HTM
Internet Chemistry Resources	http://www.rpi.edu/dept/chem/cheminfo/chemres.html
Khem Products, Inc.	http://www.khem.com/khem/home.html
Knight-Ridder Information - Dialog	http://www.dialog.com/
Nucleic Acid Database (NDB) Archive	gopher://ndbserver.rutgers.edu:70/11/etc/ndb_link_files
Online Databases, Libraries and Facilities	http://wwwchem.ucdavis.edu/chem/dbase.html
Powder Page	http://www.granular.com/
Some Chemistry Resources on the Internet--Rensselaer Polytechnic Institute	http://www.rpi.edu/dept/chem/cheminfo/chemres.html
STN International	http://www.fiz-karlsruhe.de/stn.html
Thomas	http://rs9.loc.gov/home/thomas.html
Thomas Register of American Manufacturers	http://www.thomasregister.com/

Chemical Professional and Trading Organizations

American Chemical Society (ACS)	http://www.acs.org
American Chemical Society Meetings and Conferences	http://www.acs.org/callist.htm
American Institute of Chemical Engineers Web	http://www.che.ufl.edu/aiche/
American Physical Society	http://www.aps.org/
American Vacuum Society	http://www.vacuum.org/
ASEE Clearinghouse for Engineering Education	http://www.asee.org/index.html
ATSDR Announcements	http://atsdr1.atsdr.cdc.gov:8080/
Chemical Engineering Meetings and Conferences	http://www.che.ufl.edu/www-che/index.html#announcements
Chemical Engineering Professional Organizations	http://www.che.ufl.edu/www-che/topics/professional.html
Coatings Industry Alliance	http://www.coatings.org/cia/
Electrochemical Society, Inc.	http://www.electrochem.org/ecs/
Engineering Foundation	http://www.engfnd.org/engfnd/
Gas Processors Association (GPA)	http://www.galstar.com/~gpa/
International Society of Heterocyclic Chemistry	http://euch6f.chem.emory.edu/ishc.html
Laboratory Equipment Exchange	http://www.magic.mb.ca/~econolab/
National Environmental Information Service	http://www.cais.com/tne/neis/epa_index.html

Chemical Professional and Trading Organization *(continued)*

North American Catalysis Society	http://www.dupont.com/nacs/
PolyNet	http://www.cilea.it/polynet/

Chemical Services

Advance Scientific and Chemical, Inc	http://www.sawgrass.com/advance
Advanced Visual Systems	http://www.avs.com
Aluminium Industry World Wide Web Server	http://www.euro.net/concepts/industry.html
American Chemicals Company, Inc.	http://uc.com/acci/
Analyticon Instruments Corporation	http://www.analyticon.com
Applied Coatings & Linings, Inc.	http://www.corrosion.com/applied/index.html
Aslchem International Inc	http://www.iceonline.com/home/aslchem
Baltzer Science Publishers	http://www.nl.net/~baltzer
Beilstein Information Systems	http://www.beilstein.com/
Bifurcation and Nonlinear Instability Laboratory	http://gibbs.che.ufl.edu/bifurcation.shtml
CAPD - Computer-Aided Process Design Consortium	http://www.cheme.cmu.edu/research/capd/
CAS WWW Server	http://info.cas.org/welcome.html
Ceramics and Industrial Minerals Home Page	http://www.minerals.com/~ceramics/
Chemical Education Resources	http://www.cerlabs.com/chemlabs

Chemical Services *(continued)*

Chemical Marketing Online	http://www.chemon.com/
Chemical Process Modeling and Flowsheet Synthesis	http://www.preferred.com/~lpartin/index.html
ChemSource	http://www.chemsource.com/
Chemtec Publishing	http://www.io.org/~chemtec/
CLI International, Inc.	http://www.clihouston.com/
CS Distribuidora, S.A. de C.V.	http://www.spin.com.mx/grupocs/gcs-csdb.html
DOE Office of Industrial Technologies (OIT) Chemicals Industry Team	http://www.nrel.gov/oit/Industries-of-the-Future/chemical.html
Eastern Minerals and Chemicals	http://www.ceramics.com/~ceramics/emc/
Electronic Selected Current Aerospace Notices (E-SCAN)	http://www.sti.nasa.gov/scan.html
Elsevier Science B. V.	http://www.elsevier.nl/
Finishing Industry Home Page	http://www.finishing.com/
Gilson, Inc.	http://pubs.acs.org/pin/cambrex/cam222p1.html
IndustryLink	http://www.industrylink.com/
Institute for Gas Utilization and Processing Technologies	http://www.uoknor.edu/igupt/
Interactive Simulations, Inc.	http://www.intsim.com/~isigen/
JEOL USA, Inc.	http://www.jeol.com/

Chemical Services *(continued)*

Kimberlyte Inc.	http://hookomo.aloha.net/~mikei/kimbhome.html
Low Gravity Transport Phenomena Laboratory	http://gibbs.che.ufl.edu/lowgravity.shtml
Nevada Technical Associates, Inc.	http://www.ntanet.net/
Progressive Products, Inc.	http://www.tenagra.com/progress/
Radiation Research Journal	http://www.whitlock.com/kcj/science/radres/default.htm
SAF Bulk Chemicals	http://www.sigma.sial.com/sigma/saf1hp.htm
Showa Chemical's Database	http://www.st.rim.or.jp/~shw/
SRI International	http://www.sri.com/
TeleChem International, Inc.	http://www.hooked.net/users/telechem/index.html
Thermochemical Calculator (TCC)	http://adam.caltech.edu/tcc/
Transportation Resources	http://dragon.princeton.edu:80/~dhb/
World Wide Chemnet, Inc.	http://www.galstar.com/~chemnet/index.html
WWW Chemicals	http://www3.ios.com:80/~ilyak

Chemical Supplier/Manufacturer

Aderco Fuel Additives	http://www.aderco.ca
Advanced Chemical Design	http://www.coolworld.com/shopping/advanced/index.htm
Advanced Chemicals	http://www.chataqua.com/AC/
Aerojet Chemicals	http://www3.ios.com:80/~ilyak/ind030.html
Ajax Chemicals	http://www.science.com.au/ajax/
Akzo Nobel	http://www.akzonobel.com
Amoco Corporation	http://www.amoco.com/
ARCO Chemical	http://pubs.acs.org/pin/arco/arc.html
Argus Chemicals	http://www.texnet.it/argus/argus.html
Ausimont USA, Inc.	http://Ausiusa.inter.net/ausiusa/
BASF Corporation	http://www.basf.com
Boulder Scientific Co.	http://www3.ios.com:80/~ilyak/ind019.html
Buckman Laboratories	http://www.buckman.com/
Calgon Corporation	http://www.calgon.com/
Carbolabs, Inc.	http://www3.ios.com:80/~ilyak/ind018.html
Celgene Corp.	http://www3.ios.com:80/~ilyak/ind012.html
Challenge, Inc.	http://www.challenge-inc.com/chem
Chemical Marketing Online (CHEMON)	http://www.chemon.com/
Chemnet	http://www.chemnet.com

Chemical Supplier/Manufacturer *(continued)*

Chemsyn Science Lab.	http://www3.ios.com:80/~ilyak/ind017.html
Chevron	http://www.chevron.com/
Chromophore, Inc	http://www.chromophore.com/
Chugoku Kogyo Co.,Ltd	http://chemical-metal.co.jp/cgk/
Ciba-Geigy AG Basel	http://147.167.128.11/
Clorox Company	http://www.clorox.com/
CTD, Inc	http://www.cyclodex.com/
Dalton Chemical Laboratories, Inc.	http://www.dalton.com/dalton
Deepwater Iodides, Inc.	http://www3.ios.com:80/~ilyak/ind021.html
Diaz Chemical Corp.	http://www3.ios.com:80/~ilyak/ind024.html
Dielectric Polymers Inc.	http://www.dipoly.com/
Dojindo Laboratories	http://www.dojindo.co.jp/
Dow	http://www.dow.com:80/specialty/index.html
DuPont	http://www.dupont.com
Eastman Chemical Company	http://www.eastman.com/
Eli Lilly and Company	http://www.lilly.com/
Exfluor Research Corp.	http://www3.ios.com:80/~ilyak/ind008.html
Fiberglass World	http://www.fiberglass.com/fiberglass/index.html
Fisher Scientific, Internet Catalog	http://www.fisher1.com/
FMC Corporation	http://fmcweb.ncsa.uiuc.edu/home.html

Chemical Supplier/Manufacturer *(continued)*

Furuuchi Chemical Co.	http://www.bekkoame.or.jp/~kittel/
Gelman Sciences	http://argus-inc.com/Gelman/Gelman.html
Genentech, Inc.	http://outcast.gene.com/
General Electric WWW Server	http://www.ge.com/
Goodyear Tire & Rubber Company	http://www.goodyear.com/
H&S Chemical Co.	http://www3.ios.com:80/~ilyak/ind020.html
Hampford Research, Inc.	http://www3.ios.com:80/~ilyak/ind001.html
Hampshire Chemical Corp.	http://www3.ios.com:80/~ilyak/ind003.html
Huls America Inc.	http://www3.ios.com:80/~ilyak/ind029.html
Idetec, S.A. de C.V.	http://www.spin.com.mx/grupocs/gcs-ideb.html
Indofine Chemical Co.	http://www3.ios.com:80/~ilyak/indofine.html
Interchem Corporation	http://www.interchem.com/
J.M. Huber Corporation	http://www.huber.com/
Jost Chemical	http://www3.ios.com:80/~ilyak/ind007.html
K.R. Anderson Co., Inc	http://www.kranderson.com/
Keith Ceramic Materials	http://www.ceramics.com/~ceramics/keith/
Lancaster Synthesis, Inc.	http://www3.ios.com:80/~ilyak/Lancaster.html
Lanxide Coated Products	http://www.ravenet.com/lanxcoat/

Chemical Supplier/Manufacturer *(continued)*

Materials and Electrochemical Research (MER) Corporation	http://www.opus1.com/~mercorp/index.html
Melamine Chemicals, Inc.	http://www.melamine.com/
Merck & Company, Inc.	http://www.merck.com/
Methanex	http://www.methanex.com/invest/
Millipore Corporation	http://www.millipore.com/
Millipore Corporation	http://www.millipore.com/
Misco International, Inc.	http://www.radiks.net/misco/catalog1/
Miton Products	http://www.internetpagework.com/miton.html
Mobil Corporation	http://www.mobil.com/
Monsanto Company	http://www.monsanto.com/
Morflex, Inc.	http://www3.ios.com:80/~ilyak/ind023.html
Neste Resins North America	http://www.neste-resins.com/
Norquay Technology Inc.	http://www3.ios.com:80/~ilyak/ind004.html
OM Group, Inc. (OMG)	http://www.usa.net/~omg/
Palm International	http://www.infi.net/~palm/
Papros Inc.	http://www.papros.com/
Pfizer International	http://www.cyber.nl/pfizer/
Pharm-Eco Laboratories, Inc	http://www.biospace.com/pharmeco
Pilot Chemical Company	http://www.pilotchemical.com/welcome.html

Chemical Supplier/Manufacturer *(continued)*

Pressure Chemical Co.	http://www3.ios.com:80/~ilyak/ind025.html
Quality Chemicals Inc.	http://www.firstmiss.com/qci/qci_home.html
Quimica Carnot, S.A.	http://www.spin.com.mx/grupocs/gcs-qcab.html
Reilly Industries, Inc.	http://www3.ios.com:80/~ilyak/ind005.html
Research Organics	http://resorg.com/
Rhone-Poulenc Surfactants & Specialties	http://www.rpsurfactants.com
Sachem, Inc.	http://www3.ios.com:80/~ilyak/ind022.html
Schlumberger	http://www.slb.com/
SensonCorp Ltd. (a subsididary of U.S. Tech)	http://www.usxx.com/
Shell Chemical Company	http://www.shellus.com/Chemical/Welcome.html
Sigma Chemical Company	http://www.sigma.sial.com/
Sumitomo Chemical Co.	http://www.sumitomo-chem.co.jp/
Synthetech	http://www3.ios.com:80/~ilyak/ind011.html
Texaco OnLine	http://www.texaco.com/
Toagosei Co., Ltd.	http://www.dir.co.jp/cib/4045/welcome.html
Umetrics	http://www.it-center.se/umetri/
Viscona Limited	http://www.scotborders.co.uk/horizons/viscona.html
Vista Chemical Company	http://www.vistachem.com/

EAM

Environmental Sensors	http://www.envsens.com/

Environmental Equipment Manufacturer/Supplier (Hardware)

Abitibi Environmental Technologies	http://www.industrylink.com/aet
American Turbine Pump Co	http://204.49.131.2/atp/atphome.htm
Ingot Metal Company, Ltd.	http://www.io.org/~dshore/lead.html
Lakewood Systems	http://cban.worldgate.edmonton.ab.ca/lkwd/
Molten Metal Technology	http://www.mmt.com/
Unique Tire Recycling	http://www.kpp.com/unique/unique.htm

Environmental Information Service

Alliance for Environmental Technology	http://aet.org/index.html
Alpha Analytical Labs	http://world.std.com/~alphalab/
Citylink	http://www.NeoSoft.com/citylink/
Energy & Environmental Research Center	http://www.eerc.und.nodak.edu/
Energy Federation, Inc. (EFI)	http://www.tiac.net/users/efi/
EnviroLink Network	http://envirolink.org/start_web.html
Enviromine	http://www.infomine.com/technomine/enviromine/env_main.html
Environmental Library	http://envirolink.org/envirowebs.html

Environmental Information Service *(continued)*

Environmental Professional's Guide to the Net (EPGN)	http://www.geopac.com/
Environmental Protection Agency WWW Server	http://www.epa.gov/
Enviroweb	http://envirolink.org/start_web.html
EnviroWeb -- A Project of the EnviroLink Network	http://envirolink.org/
Friends of Earth Home Page	http://www.foe.co.uk
IndustryNET	http://www.industry.net/
National Institute of Environmental Health Sciences	http://lmb.niehs.nih.gov/home.html
National Technology Transfer Center (NTTC)	http://iridium.nttc.edu/nttc.html
Publishers	http://www.comlab.ox.ac.uk/archive/publishers.html
WAU Process Engineering: Marine and Environmental Biotechnology	http://www.spb.wau.nl/prock/rene.html
World-Wide Web Virtual Library: Environmental Engineering	http://www.nmt.edu/~jjenks/engineering.html
World-Wide Web Virtual Library: Wastewater Engineering	http://www.halcyon.com/cleanh2o/ww/welcome.html

Environmental Professional and Trading Organization

American Public Works Association	http://www.pubworks.org/apwa/main.html
AWWA (American Water Works Association)	http://www.awwa.org/
NAEP (National Association of Environmental Professionals)	http://enfo.com/NAEP

Environmental Protection Agency

EPA Acid Rain Hotline	http://www.epa.gov/access/chapter3/s1-1.html
EPA Air Risk Information Support Center Hotline	http://www.epa.gov/access/chapter3/s1-2.html
EPA Alternative Treatment Technology Information Center	http://www.epa.gov/access/chapter3s1-13.html
EPA Asbestos Ombudsman Clearinghouse/Hotline	http://www.epa.gov/access/chapter3/s3-1.html
EPA Clean Lakes Clearinghouse	http://www.epa.gov/access/chapter3/s3-14.html
EPA Clean-Up Information Bulletin Board System	http://www.epa.gov/access/chapter3/s1-14.html
EPA Clearinghouses and Hotlines	http://www.epa.gov/
EPA Control Technology Center	http://www.epa.gov/access/chapter3/s1-3.html
EPA Emergency Planning and Community Right-to-Know Information Hotline	http://www.epa.gov/access/chapter3/s1-15.html
EPA Emission Factor Clearinghouse	http://www.epa.gov/access/chapter3/s1-4.html

Environmental Protection Agency *(continued)*

EPA Environmental Financing Information Network	http://www.epa.gov/access/chapter3/s3-23.html http://www.epa.gov/access/chapter3/s3-25.html
EPA Green Lights Program	http://www.epa.gov/access/chapter3/s1-6.html
EPA Hazardous Waste Ombudsman Program	http://www.epa.gov/access/chapter3/s1-16.html
EPA Indoor Air Quality Information Clearinghouse	http://www.epa.gov/access/chapter3/s1-7.html
EPA INFOTERRA	http://www.epa.gov/access/chapter3/s2-1.html
EPA Institute	http://www.epa.gov/access/chapter3/s3-24.html
EPA International Cleaner Production Information Clearinghouse	http://www.epa.gov/access/chapter3/s3-9.html
EPA Methods Information Communications Exchange	http://www.epa.gov/access/chapter3/s1-17.html
EPA Model Clearinghouse	http://www.epa.gov/access/chapter3/s1-5.html
EPA National Air Toxics Information Clearinghouse	http://www.epa.gov/access/chapter3/s1-8.html
EPA National Lead Information Center Hotline	http://www.epa.gov/access/chapter3/s3-2.html
EPA National Pesticide Information Retrieval System	http://www.epa.gov/access/chapter3/s3-3.html
EPA National Pesticide Telecommunications Network	http://www.epa.gov/access/chapter3/s3-4.html

Environmental Protection Agency *(continued)*

EPA National Radon Hotline	http://www.epa.gov/access/chapter3/s1-9.html
EPA National Response Center	http://www.epa.gov/access/chapter3/s1-18.html
EPA National Small Flows Clearinghouse	http://www.epa.gov/access/chapter3/s3-15.html
EPA Nonpoint Source Information Exchange	http://www.epa.gov/access/chapter3/s3-16.html
EPA Office of Air Quality Planning and Standards Technology Transfer Network Bulletin Board System	http://www.epa.gov/access/chapter3/s1-10.html
EPA Office of Research and Development Electronic Bulletin Board System	http://www.epa.gov/access/chapter3/s3-13.html
EPA Office of Water Resource Center	http://www.epa.gov/access/chapter3/s3-17.html
EPA OzonAction	http://www.epa.gov/access/chapter3/s3-10.html
EPA Pollution Prevention Information Clearinghouse	http://www.epa.gov/access/chapter3/s3-11.html
EPA Pollution Prevention Information Exchange System	http://www.epa.gov/access/chapter3/s3-12.html
EPA Reasonably Available Control Technology, Best Available Control Technology, and Lowest Achievable Emission Rate Clearinghouse	http://www.epa.gov/access/chapter3/s1-11.html
EPA Resource Conservation and Recovery Act/Superfund/Underground Storage Tank Hotline	http://www.epa.gov/access/chapter3/s1-19.html

Environmental Protection Agency *(continued)*

EPA Risk Communication Hotline	http://www.epa.gov/access/chapter3/s3-28.html
EPA Safe Drinking Water Hotline	http://www.epa.gov/access/chapter3/s3-18.html
EPA Solid Waste Assistance Program	http://www.epa.gov/access/chapter3/s1-20.html
EPA Stratospheric Ozone Information Hotline	http://www.epa.gov/access/chapter3/s1-12.html
EPA The 33/50 Program, Special Projects Office (SPO), Office of Pollution Prevention and Toxics	http://www.epa.gov/access/chapter3/s3-5.html
EPA Toxic Release Inventory User Support	http://www.epa.gov/access/chapter3/s3-6.html
EPA Toxic Substances Control Act Assistance Information Service	http://www.epa.gov/access/chapter3/s3-7.html
EPA Wastewater Treatment Information Exchange	http://www.epa.gov/access/chapter3/s3-20.html

Environmental Services

AGRA Earth and Environmental Limited	http://www.kingsu.ab.ca/~agra/agra.html
Alberta Research Council	http://skyler.arc.ab.ca/ARC-research.html
ASL - Analytical Service Laboratories	http://www.asl-labs.bc.ca/
Augias Environmental Corp.	http://www.pronett.com/augias/augias.htm
Bio Control Network	http://www.usit.net/BICONET
Communicopia Environmental Research and Communications	http://www.communicopia.bc.ca

Environmental Services *(continued)*

EMAX Solution Partners	http://www.emax.com/
Environment One Corporation	http://www.eone.com/eone/
Environmental News Network	http://www.enn.com
ETS International, Inc.	http://www.infi.net/~etsas/
Farrell Research	http://www.ionet.net/~enviro/enviro.shtml
Geraghty & Miller, Inc.	http://www.gmgw.com/gm
Global Network of Environment and Technology	http://www.gnet.org/
Green Market	http://www.igc.apc.org/GreenMarket/
Greenfield Environmental	http://www.greenfield.com/
GreenSoft Corporation	http://www.greendesk.com/
Hydrocomp, Inc.	http://www.hydrocomp.com/
Indoor Air Quality (IAQ) Publications	http://www.iaqpubs.com/
INTERA Inc.	http://www.nmia.com/~interags/home/home.html
Mittelhauser	http://www.mittelhauser.com
Morris Environmental	http://www.morrisenv.com/
Nutting Environmental of Florida	http://www.gate.net/~nutting/
Rescan International Inc.	http://www.rescan.com/res/
Resource Development Associates	http://www.sonic.net/~rossand/page.html
Rhyme Industries	http://www.envirolink.org/products/rhyme
SCC Environmental	http://www.compusult.nf.ca/~pantle/

Environmental Services *(continued)*

Stormceptor,Inc.	http://www.stormceptor.ca
TSD Central (Hazardous Waste Central)	http://www.tsdx.com/central
WMX Technologies	http://www.wmx.com/

Environmental Wastewater Information Service

Environmental Industry Web Site	http://www.enviroindustry.com/
Finishing.com	http://www.finishing.com/index.html
Hazardous Materials Management	http://www.io.org/~hzmatmg/
IndustryNET	http://www.industry.net/
Membrane Technology Group Via-web:	http://utct1029.ct.utwente.nl/documents/membrane.html
Solutions Software	http://www.env-sol.com/
TechExpo (TM)	http://www.techexpo.com
Water - Wastewater Web	http://www.w-ww.com
WaterOnLine	http://www.wateronline.com/
Waterweb	http://www.waterweb.com
Waterworld	http://www.waterworld.com/
WWTNET	http://users.aol.com/wwtnet/wwtnet.html

Environmental Wastewater Treatment Equipment Manufacturer/Supplier

ABS (Aerated Biological Surfaces) Inc.	http://execpc.com/~abs/
Calgon Corporation	http://www.calgon.com/
DMP Corporation	http://web.sunbelt.net/dmp/dmp.htm
ECI (Environmental Concerns, Inc.)	http://www.datacor.com/~eci
National Filter Media Corp.	http://www.xmission.com/~nfm/
Park Equipment Company	http://rampages.onramp.net/~parkco/
Remco Engineering	http://www.remco.com./~remcobob/home.htm
Trojan Technologies Inc.	http://www.trojanuv.com/
Universal Enviromental Technologies	http://www.uet.com/

Environmental Wastewater Treatment Facilities (Municipal)

Evaluation of Artificial Wetlands	http://gus.nsac.ns.ca/~piinfo/resman/wetlands/anno/annobib.html
Foundation for Cross-Connection Control & Hydraulic Research	http://www.usc.edu:80/dept/fccchr/
The Frank E. VanLare Wastewater Treatment Facility	http://www.history.rochester.edu/class/vanlare/home.htm
Sewage in Puget Sound	gopher://futureinfo.com/1/menu3/menu5/
Utility, Power, Wastewater Plant's E-mail Directory	http://www2.best.com/~lidial/plant.htm
Water & Wastewater Utilities	http://www.waterweb.com/ww/utility.html

Environmental Wastewater Treatment Services

Black & Veatch	http://www.bv.com/
CH2M Hill	http://www.ch2m.com/
Greenspan Technology: Environmental Monitoring	http:/peg.pegasus.oz.au/~greenspan/
Hydromantis	http://www.hydromantis.com/index.html
Reid Engineering	http://users.aol.com/reidengnr/reid.html
Rothberg, Tamburini & Winsor, Inc.	http://204.131.74.102/RTW/default.htm
SCS Engineers	http://204.240.184.66/index.html
Utah Water Research Laboratory	http://publish.uwrl.usu.edu/uwrl.gome.html
Wastewater Solutions	http://pages.prodigy.com/MO/wastewater/company.html
Water Resources Management, Inc.	http://www.wrmi.com/pub/wrmi/wrmi.htm

Government

Agency for Toxic Substances and Disease Registry (ATSDR)	http://atsdr1.atsdr.cdc.gov:8080/
Agency for Toxic Substances and Disease Registry (ATSDR) Science Corner	http://atsdr1.atsdr.cdc.gov:8080/cx.html
Carbon Dioxide Information Analysis Center (CDIAC)	http://cdiac.esd.ornl.gov
Center of Disease Control (CDC and Prevention)	http://www.cdc.gov
Department of Commerce (DOC)	http://www.doc.gov
Department of Energy (DOE)	http://www.doe.gov
Department of Labor (DOL)	http://www.dol.gov
Environmental Protection Agency (EPA)	http://www.epa.gov/
Federal Printing Office	http://www.access.gpo.gov
FedWorld Information Network	http://www.fedworld.gov/index.html
Food and Drug Administration (FDA)	http://www.fda.gov
Food and Drug Administration (FDA) Via-telnet:	fdabbs.fda.gov; login: bbs; password: bbs
Library of Congress (LOC)	http://www.loc.gov
National Institute of Environmental Health (NIEHS)	http://www.niehs.nih.gov
National Institute of Health (NIH)	http://www.nih.gov
National Institute of Occupational Safety and Health (NIOSH)	http://www.cdc.gov/niosh/homepage.html

Government *(continued)*

National Institute of Standards and Technology (NIST)	http://www.nist.gov
National Library of Medicine (NLM)	http://www.nlm.nih.gov
National Oceanic and Atmospheric Administration (NOAA)	http://www.noaa.gov
National Oceanic and Atmospheric Administration (NOAA) Environmental Information Services	http://www.esdim.noaa.gov
National Science Foundation	http://www.nsf.gov
National Technical Information Service (NTIS)	http://www.fedworld.gov/ntis/ntishome.html
Occupational Safety and Health Administration (OSHA)	http://www.osha.gov

Life Cycle Analysis

LCA at the University of Toronto	http://www.ecf.toronto.edu/~young/

P2 Information Service

Basel Convention	http://www.unep.ch/sbc/about.html
Burton Hamner's list of Internet Environmental Sources	http://ie.uwindsor.ca/ecdmlist/feb1995.2.html#feb179503
Center for Clean Technology	http://cct.seas.ucla.edu/cct.pp.html
Communications for a Sustainable Future	http://csf.colorado.edu/
Consortium on Green Design and Manufacturing	http://euler.berkeley.edu/green/cgdm.html

P2 Information Service *(continued)*

Design for the Environment	http://w3.pnl.gov:2080/DFE/home.html
ECDM Group	http://www.me.mtu.edu/research/envmfg/
Ecocycle	http://www.doe.ca/ecocycle/
Energy and Environmentally Conscious Manufacturing	http://www.ornl.gov/orcmt/energy/home.html
Enviroene	http://wastenot.inel.gov/envirosense/
Environment Canada	http://www.doe.ca/
Environmental Industry Web Site	http://www.enviroindustry.com/
Environmentally Conscious Design and Manufacturing Lab	http://ie.uwindsor.ca/ecdm_lab.html
EnviroWeb	http://envirolink.org/
EPA Wastewi$e Program	http://cygnus-group.com/ULS/Waste/epa.html
Green Design Initiative	http://www.ce.cmu.edu:8000/GDI/
Industrial Assessment Center, University of Florida	http://www.che.ufl.iac
Information Technology for Environmentally Conscious Design, Construction and Manufacturing	http://iv.cee.tufts.edu:8000/berger_chair.html
Integrating Environment and Development	http://www.erin.gov.au/portfolio/esd/integ.html
Interduct	http://dutw239.tudelft.nl
International Organization for Standardization (ISO)	http:/www.iso.ch/
ISO 14000 Information	http://www.stoller.com/iso.htm

P2 Information Service *(continued)*

IVAM Environmental Research	http://www.ivambv.uva.nl/welcome.html
Materials Systems Laboratory	http://web.mit.edu/org/c/ctpid/www/msl/index.html
Montreal Protocol on Substances that Deplete the Ozone Layer	http://www.greenpeace.org/~intlaw/mont-htm.html
National Key Centre for Design	http://daedalus.edc.rmit.edu.au/
National Pollution Prevention Center for Higher Education	http://www.snre.umich.edu/nppc/
O2 Global Network	http://www.wmin.ac.uk/media/O2/O2_Home.html
Office of Industrial Productivity and Energy Assessment	http://oipea-www.rutgers.edu/html_docs/waste&p2.html
Office of Pollution Prevention and Compliance Assistance	http://www.dep.state.pa.us/dep/deputate/pollprev/iso14000/isopart.htm
Pollution Prevention and Waste Minimization	http://www.sme.org/apaa/pollut.html
Pollution Prevention Program Database Via-gopher:	gopher://gopher.pnl.gov:2070/1/.pprc
Reverse Logistics	http://www.tue.nl/ivo/lbs/main/index.htm
Scott Butner's list of Internet Environmental Sources	http://ie.uwindsor.ca/ecdm_info/butner.html
SimaPro3	http://www.ivambv.uva.nl/pre/simapro.html
Solstice	http://solstice.crest.org/
Sustainable Development Online	http://www.iso.ch/
Systems Realization Laboratory	http://www.srl.gatech.edu/

P2 Information Service *(continued)*

Technology, Business and the Environment Program	http://web.mit.edu/afs/athena/org/c/ctpid/www/tbe/index.html
Tres English's Sustainable Development pages	http://www.azstarnet.com/~tres/
UCLA CCT Home Page	http://cct.seas.ucla.edu/

PLI

Plastics Network	http://www.plasticsnet.com/index.html
PolyLinks	http://www.polymers.com/
PolySort	http://www.polysort.com

PLM

AlliedSignal	http://www.polysort.com/allied
Associated Rubber Company	http://www.mindspring.com/~asscrc/
Dielectric Polymers Inc.	http://www.dipoly.com/
Engineered Rubber Products	http://www.polysort.com/ERP
General Electric (GE) Plastics	http://www.ge.com/gep/homepage.html
HAAKE	http://www.polysort.com/haake
ICI Fiberite	http://www.olworld.com./olworld/mall/mall_us/c_busfin/m_fiberi/
Industrial Plastics and Paints	http://storefront.net/storefront/ipp/index.html
Industrial Services International (Terra-Sorb)	http://www.cais.com/cytex/isi/tsorb.html

PLM *(continued)*

Park Scientific Instruments	http://www.park.com/
Phoenix Polymers, Inc.	http://www.iii.net/biz/phoenix.html
Quadrax Corp.	http://www.growth.com/MENU/QDRX/QDRXhome.html
RJF International Corporation	http://www.polysort.com/RJF
Rohm and Haas Company	http://www.rohmhaas.com/
Struktol Company of America	http://www.polysort.com/companies/s/struktol/ strkhome.htm
U.S. Technology	http://www.polysort.com/ustech
Van Dorn Demag	http://www.polysort.com/vandorn
Zefon Manufacturing	http://www.zefon.com/
Zipperling Kessler & Co	http://www.zipperling.de/

PLS

Akromold Inc.	http://www.polysort.com/akromold
American Mold & Engineering	http://AmericanMold.com/~ame/
ASB Industries	http://www.polysort.com/ASB
ATI Cahn Company	http://www.netopia.com:80/aticahn/
Atlantis Plastics, Inc.	http://www.cfonews.com/agh/
Custom Plastic Extrusions	http://www.polysort.com/companies/c/cpe/cpe.html
E & D Plastics	http://edp.ncc.com/edp/
Fast Heat	http://www.ais.net/fastheat/

PLS *(continued)*

General Plastex	http://www.polysort.com/gplastex
Gilchrist Polymer Center	http://www.polysort.com/gilchrist
Harbec Plastics	http://www.harbec.com/
Husky Injection Molding Systems	http://www.husky.on.ca/
Integrated Design Engineering Systems (IDES)	http://www.idesinc.com/
Jamestown Tooling & Machining	http://www.jtm.com/
Lauren Manufacturing	http://www.polysort.com/lauren
MacroFAQS	http://www.polysort.com/FAQS
Maxima Plastics	http://www.net-link.net/maxima
Metcalfe Plastics Corporation	http://www.metcalfe.com
Moldflow Ltd	http://www.worldserver.pipex.com/moldflow/
Mother Lode Plastics	http://www.sonnet.com/webworld/mplastic.htm
Multibase	http://www.polysort.com/multibase
Ohio Valley Plastics Partnership	http://www.polysort.com/ohiovaly
Old Line Plastics	http://www.charm.net/~olp/
Parr-Green	http://www.polysort.com/parrgreen
Performance Plastics	http://www.polysort.com/PPI
Plasti Dip Plastic Spray	http://www.cais.com/cytex/pdi/pdi.html
Plastic Express	http://www.polysort.com/px

PLS *(continued)*

Plastic Technology Center	http://www.lexmark.com/ptc/ptc.html
The Plastics Group	http://www.theplastics.com/tpg
Plastics Hotline	http://www.polysort.com/plasthot
Polaris Plastic Sales	http://www.polysort.com/polaris
Pro-Mold	http://www.polysort.com/promold
QMR Plastics	http://www.spacestar.com/users/barqmr/
Replas	http://www.replas.com/
Reuther Mold & Machine	http://www.polysort.com/reuther
Ridout Plastics - Award Winning Plastics Company	http://www.sddt.com/~plastics/
Ro-Mai Industries	http://www.polysort.com/RMI
Robert Eller Associates	http://www.polysort.com/REA
Sajar Plastics	http://www.polysort.com/sajar
Smith's Enterprises	http://www.iquest.net/~ariel/ibd/label13.html
Spirex	http://www.polysort.com/spirex
STN International--the Scientific and Technical Information Network	http://www.fiz-karlsruhe.de/pdlcom.html
Tech Mold	http://www.industrylink.com/techmold
Technology and Trade Inc.	http://www.techtrade.com
Universal Plastics	http://www.polysort.com/universal
VPI	http://www.dataplusnet.com/vpi/vpi.html
Witt Plastics	http://lakeland.tsolv.com/~wittpl/

Recycling

Automotive Recycling Mailing List	http://ie.uwindsor.ca/autorecy/welcome.html
ChemSearch (Chemical Recycling)	http://www.sonic.net/chemsearch
Chrysler Corp. Recycling	http://www.chryslercorp.com/environment/recycling.html
Eco-Glass Group	http://www.recycle.net/recycle/Trade/rs000749.html
Environmental Recycling Hotline	http://www.primenet.com/erh.html
Environmental Technologies USA, Inc.	http://www.mvisibility.com/ENVR/
The Exchange	http://www.earthcycle.com/g/p/earthcycle//
Ford Motor Co. Recycling	http://www.ford.com/corporate-info/environment/Recycling.html
FRC International	http://norden1.com/~frc/
Global Recycling Network, Inc.	http://grn.com/grn/
Lackie & Associates (Recycler's World)	http://www.recycle.net/recycle/Trade/rs000811.html
Plastic Bag Association	http://penbiz.com/stellar/green/
Recycler's World	http://www.recycle.net/recycle/index.html
Recycling WWW sites	http://ie.uwindsor.ca/ecdm_info/recycle.html
Ribbon-Jet Tek	http://usa.net/ca/casual.html
Specialty Recycling Services	http://www.polysort.com/SRS
TreeEco	http://www.envirolink.org/treeco/

World Wide Web Engine

Aliweb	http://www.traveller.com/aliweb/
Alta Vista	http://www.altavista.digital.com/
EINET Galaxy	http://galaxy.einet.net/
Inktomi	http://inktomi.berkeley.edu/

World Wide Web Virtual Library

Chemical Engineering	http://www.che.ufl.edu/www-che/
Chemical Engineering: Information Indexes	http://www.che.ufl.edu/www-che/topics/indexes.html
Chemical Engineering: Professional Organization	http://www.che.ufl.edu/www-che/topics/research.html
Chemical Engineering: Research Organization and Laboratories	http://www.che.ufl.edu/www-che/topics/research.html
Chemistry	http://www.chem.ucla.edu/chempointers.html
Chemistry Sites at Commercial Organizations	http://www.chem.ucla.edu/chempointers.html#www_commercial
Chemistry Sites at Non-Profit Organizations	http://www.chem.ucla.edu/chempointers.html#www_nonprofit
Chemistry Sites at Other Information Resources	http://www.chem.ucla.edu/chempointers.html#www_vl
Civil Engineering	http://www.ce.gatech.edu/WWW-CE/home.html
Environmental Engineering	http://www.nmt.edu/~jjenks/enfginering.html

World Wide Web Virtual Library *(continued)*

Green Engineering	http://ie.uwindsor.ca/other_green.html
Wastewater Engineering	http://www.halcyon/cleanh2o/ww/
Wastewater Information	http://www.halcyon.com/cleanh2o/ww/wwtinfo.html

CHAPTER 3

ACADEMIC INSTITUTIONS

Chapter 3

Academic Institutions

Chemical Engineering Programs

Arizona State University, College of Engineering and Applied Sciences, Chemical Engineering Department	http://www.eas.asu.edu/
Auburn University, College of Engineering, Department of Chemical Engineering	http://www.eng.auburn.edu/ department/che/chehome.html
Brigham Young University, College of Engineering and Technology, Department of Chemical Engineering	http://www.et.byu.edu/centers/ cheme-info.html
Brown University, Division of Engineering, Chemical Engineering Program	http://sunserver.engin.brown. edu:6666/
Bucknell University, College of Engineering, Chemical Engineering Department	http://spectrum.eg.bucknell.edu/ cm.html
California Institute of Technology, Division of Chemistry and Chemical Engineering	http://www.che.caltech.edu/
California State University, Long Beach, College of Engineering, Chemical Engineering Department	http://www.engr.csulb.edu/chem/
California State University, Sacramento, School of Engineering and Computer Science, Chemical and Materials Engineering Department	http://hera.ecs.csus.edu/
Carnegie Mellon University, Department of Chemical Engineering	http://www.cheme.cmu.edu/

Chemical Engineering Programs *(continued)*

Case Western Reserve University, Case School of Engineering, Department of Chemical Engineering	http://k2.scl.cwru.edu/cse/eche/
Christian Brothers University, School of Engineering, Department of Chemical Engineering	http://www.cbu.edu/engineering/chedept.htm
City College of the City University of New York (The), Chemical Engineering Department	http://www.cuny.edu/
Clarkson University, Department of Chemical Engineering	http://fire.clarkson.edu/
Clemson University, College of Engineering, Department of Chemical Engineering	http://macmosaic.eng.clemson.edu/Academic.Depts/ChemE/ChE.html
Cleveland State University, Chemical Engineering Department	http://www.csuohio.edu/
Colorado School of Mines, Division of Engineering, Chemical Engineering Department	http://www.mines.colorado.edu:7024/0/index.html
Colorado State University, Chemical Engineering Department	http://stokes.lance.colostate.edu/
Columbia University, School of Engineering and Applied Science, Department of ChemicalEngineering, Materials Science and Mining Engineering	http://www.seas.columbia.edu/columbia/departments/chemical_engineering/
Cooper Union, The Albert Nerken School of Engineering, Department of Chemical Engineering and Chemistry	http://www.cooper.edu/engineering/chemechem/Welcome.html

Chemical Engineering Programs *(continued)*

Cornell University, College of Engineering, School of Chemical Engineering	http://www.cheme.cornell.edu/
Dartmouth College, Thayer School of Engineering, Biotechnology and Biochemical Engineering Program	http://caligari.dartmouth.edu/thayer/
Drexel University, College of Engineering, Chemical Engineering Department	http://www.drexel.edu/Engineering.html
Florida Agriculture and Mechanical University/Florida State University, Department of Chemical Engineering	http://www.eng.fsu.edu/home_pages/vinals/cheme.html
Florida Institute of Technology, College of Engineering, Chemical Engineering Program	http://sci-ed.fit.edu/engsci/chemical/chemical.html
Georgia Institute of Technology, School of Chemical Engineering	http://www.chemse.gatech.edu/
Hampton University, School of Engineering & Technology, Chemical Engineering Program	http://137.198.30.10//cur/cme.html
Harvey Mudd College of Engineering and Science Engineering Department, Chemical Engineering Department	http://www.hmc.edu/acad/eng/
Howard University, School of Engineering, Department of Chemical Engineering	http://www.cldc.howard.edu/~haydeng/
Illinois Institute of Technology, Armour College of Engineering, Chemical Engineering Department	http://www.iit.edu/dept/che/

Chemical Engineering Programs *(continued)*

Institute of Paper Science and Technology	http://www.ipst.edu
Iowa State University of Science and Technology, College of Engineering, Department of Chemical Engineering	http://www.public.iastate.edu/~cheme/homepage.html
Johns Hopkins University, The Whiting School of Engineering, Chemical Engineering Department	http://www.jhu.edu/~cheme/ChemE.html
Kansas State University, Department of Chemical Engineering	http://www.engg.ksu.edu/CHEDEPT/home.html
Lafayette College, Chemical Engineering Department	http://www.lafayette.edu/rosenbau/eng_html/cheme.htm
Lamar University, College of Engineering, Chemical Engineering Department	http://www.lamar.edu/lamar/chemeng.html
Lehigh University, College of Engineering and Applied Science, Chemical Engineering Department	http://www.lehigh.edu/~ak04/ak04.html
Louisiana State University, Department of Chemical Engineering	http://svr1.che.lsu.edu/
Louisiana Tech University, Chemical Engineering Department	http://aurora.latech.edu/
Manhattan College, School of Engineering, Chemical Engineering Department	http://www.cc.mancol.edu/engineer/chmlpage.html
Massachusetts Institute of Technology, Department of Chemical Engineering	http://web.mit.edu/afs/athena/org/c/cheme/www/Titlepage.html
McNeese State University, Chemical and Electrical Engineering Department	http://www.mcneese.edu/

Chemical Engineering Programs *(continued)*

Michigan State University, College of Engineering, Department of Chemical Engineering	http://www.egr.msu.edu/ChE/
Michigan Technological University, Division of Chemical Sciences, Chemical Engineering Department	http://www.chem.mtu.edu/
Mississippi State University, College of Engineering, Chemical Engineering Department	http://www.de.msstate.edu/
Montana State University-Bozeman, College of Engineering, Chemical Engineering Department	http://www.coe.montana.edu/che/
New Jersey Institute of Technology, Newark College of Engineering, Chemical Engineering Department	http://www.njit.edu/njIT/Schools/NCE/chem.html
New Mexico State University, College of Engineering, Chemical Engineering Department	http://chemeng.nmsu.edu/
North Carolina Agricultural and Technical State University, School of Engineering, Chemical Engineering Department	http://www.ncat.edu/engineer/chem/Welcome.html
North Carolina State University at Raleigh, Department of Chemical Engineering	http://www.eos.ncsu.edu/coe/departments/che/index.html
Northeastern University, College of Engineering, Chemical Engineering Department	http://www.northeastern.edu/registrar/catalog/che.html

Chemical Engineering Programs *(continued)*

Northwestern University, McCormick School of Engineering and Applied Science, Department of Chemical Engineering (*see also* McCormick Graduate Study in Chemical Engineering).	http://www.chem-eng.nwu.edu/ (*See also* http://www.chem-eng.nwu.edu)
Ohio State University, The Chemical Engineering Department	http://kcgl1.eng.ohio-state.edu/che/home_page.html
Ohio University, Russ College of Engineering and Technology, Department of Chemical Engineering	http://www.ent.ohiou.edu/che/
Oklahoma State University, College of Engineering, Architecture, and Technology, School of Chemical Engineering	http://www.okstate.edu/ceat/chemeng/
Oregon State University, Department of Chemical Engineering	http://www.engr.orst.edu/~reed/CSTR/index.html
Pennsylvania State University, The Department of Chemical Engineering	http://www.engr.psu.edu/www/dept/che/ index.html
Polytechnic University (Brooklyn), Academic Departments, Chemical Engineering, Chemistry and Material Science Department	http://www.poly.edu/polytech/academia/home.html
Prairie View A&M University College of Engineering and Architecture, Chemical Engineering Department	http://pvcea.pvamu.edu/
Princeton University, Department of Chemical Engineering	http://arnold.princeton.edu/
Purdue University, School of Chemical Engineering	http://che.www.ecn.purdue.edu/

Chemical Engineering Programs *(continued)*

Rensselaer Polytechnic Institute, Department of Chemical Engineering (includes public domain educational materials for biochemical and environmental engineering courses).	http://www.eng.rpi.edu:80/dept/chem-eng/teach.html
Rice University, Department of Chemical Engineering	http://www.ruf.rice.edu/~che/
Rose-Hulman Institute of Technology, Chemical Engineering Department	http://www.rose-hulman.edu/
Rutgers University (The), The State University of New Jersey, Chemical and Biochemical Engineering Graduate Program	http://sol.rutgers.edu/
San Jose State University, College of Engineering, Chemical Engineering Department	http://130.65.150.50/cheme.htm
South Dakota School of Mines and Technology, Department of Chemistry and Chemical Engineering	http://www.sdsmt.edu/campus/chem/chem.html
Stanford University, School of Engineering, Chemical Engineering Department	http://www-leland.stanford.edu/group/chemeng/
State Technical Institute of Memphis, Chemical Engineering Technology	gopher://stim.tec.tn.us/1
State University of New York at Buffalo, Chemical Engineering Department	http://www.eng.buffalo.edu/dept/ce/
Stevens Institute of Technology, Chemistry and Chemical Engineering Department	http://www.stevens-tech.edu/stevens/chem/chem.html

Chemical Engineering Programs *(continued)*

Syracuse University, L.C. Smith College of Engineering and Computer Science, Chemical Engineering and Materials Science	http://www.ecs.syr.edu/
Texas A&M University - Kingsville, College of Engineering, Chemical and Natural Gas Engineering Department	http://mullet.taiu.edu/CHENGEHome.html
Texas A&M University, Chemical Engineering Department	http://www-chen.tamu.edu/
Texas Tech University, College of Engineering, Chemical Engineering Department	http://www.coe.ttu.edu/che/che_home.htm
Tufts University, College of Engineering, Chemical Engineering Department	http://www.tufts.edu/as/engdept/
Tulane University, Chemical Engineering Department	http://www.tulane.edu/~bmitche/index.html
University of Akron, Chemical Engineering Department	http://ecgfhp01.ecgf.uakron.edu/Chem.html
University of Alabama, College of Engineering, Chemical Engineering Department	http://hamton.eng.ua.edu/college/che.html
University of Alabama, Huntsville, College of Engineering, Department of Chemical and Materials Engineering	http://eb-p5.eb.uah.edu/che/general/che_home.html
University of Arizona, College of Engineering and Mines, Department of Chemical and Environmental Engineering	http://www.che.arizona.edu

Chemical Engineering Programs *(continued)*

University of Arkansas, Fayetteville, Chemical Engineering Department (includes a link to the Robert N. Maddox Chemical Engineering Technical Reference Center Card Catalog).	http://www.engr.uark.edu/engr/departments/cheg/card_cat.html
University of California, Riverside, The Marlan and Rosemary Bourns College of Engineering, Chemical Engineering Department	http://engr.ucr.edu/chem/chem_base.html
University of California, Los Angeles, School of Engineering and Applied Science, Chemical Engineering Department	http://www.seas.ucla.edu/ch/
University of California, Santa Barbara, College of Engineering, Department of Chemical Engineering	http://eci.ucsb.edu/ce/
University of California, San Diego, School of Engineering, Applied Mechanics and Engineering Sciences Department	http://www-ames.ucsd.edu/
University of California, Irvine, School of Engineering, Department of Chemical and Biochemical Engineering	http://www.eng.uci.edu/cbe/
University of California, Berkeley, College of Chemistry, Department of Chemical Engineering	http://www.cchem.berkeley.edu/ChemE/index.html
University of California, Davis, College of Engineering, Department of Chemical Engineering and Materials Science	http://nachos.engr.ucdavis.edu/~chmsweb/
University of Cincinnati, College of Engineering, Chemical Engineering Department	http://www.eng.uc.edu/che/

Chemical Engineering Programs *(continued)*

University of Colorado, Boulder, College of Engineering and Applied Science, Chemical Engineering Department	http://coll-engr.colorado.edu/Profile/Chemical.html
University of Connecticut, School of Engineering, Chemical Engineering Department	http://www.eng2.uconn.edu/cheg/index.html
University of Dayton, School of Engineering, Chemical and Materials Engineering Department	http://www.engr.udayton.edu/SOE/Depts/chemical/chemical.htm
University of Dayton, School of Engineering, Department of Chemical Process Technology	http://www.engr.udayton.edu/SOE/Depts/chemtech/chemtech.htm
University of Delaware, Chemical Engineering Department	http://www.che.udel.edu/
University of Detroit, Mercy, Chemical Engineering Science Department	http://www.udmercy.edu/
University of Florida, College of Engineering, Chemical Engineering Department	http://www.che.ufl.edu
University of Houston, Chemical Engineering Department	http://www.egr.uh.edu/CHEE/Welcome.html
University of Idaho, College of Engineering, Chemical Engineering Department	http://www.uidaho.edu/che/
University of Illinois, Chicago, College of Engineering, Chemical Engineering Department	http://www.uic.edu/depts/chme/
University of Illinois at Urbana-Champaign, College of Engineering, Department of Chemical Engineering	http://www.scs.uiuc.edu:80/chem_eng/

Chemical Engineering Programs *(continued)*

University of Iowa, College of Engineering, Chemical Engineering Department	http://www.icaen.uiowa.edu/
University of Kansas, College of Engineering, Department of Chemical and Petroleum Engineering	http://www.engr.ukans.edu/cpe-grad/
University of Kentucky, College of Engineering, Department of Chemical and Materials Engineering	http://www.engr.uky.edu/CME/CMEhome.html
University of Louisville Speed Scientific School, Chemical Engineering Department	http://www.spd.louisville.edu/chemical/chemical.html
University of Maine, College of Engineering, Department of Chemical Engineering	http://www.umecheme.maine.edu/che/
University of Maryland, Baltimore County, College of Engineering, Department of Chemical and Biological Engineering	http://echo.umd.edu/schools/umd/dept/chemical/chem.html
University of Massachusetts at Lowell, The James B. Francis College of Engineering, Chemical Engineering Department	http://www.uml.edu/Colleges/Engineering.html
University of Massachusetts at Amherst, Department of Chemical Engineering	http://www.ecs.umass.edu/che/
University of Michigan, Ann Arbor, Department of Chemical Engineering	http://www.engin.umich.edu/labs/mel/Chemengin
University of Minnesota, Department of Chemical Engineering and Materials Science	http://www.cems.umn.edu/

Chemical Engineering Programs *(continued)*

University of Mississippi, School of Engineering, Department of Chemical Engineering	http://sunset.backbone.olemiss.edu/depts/chemical_eng/
University of Missouri-Rolla, Chemical Engineering Department	http://www.eng.umr.edu/WWWENG/CHE.html
University of Missouri-Columbia, College of Engineering, Department of Chemical Engineering	http://www.ecn.missouri.edu/academic/chem/index.html
University of Nebraska, Lincoln, Chemical Engineering Department	http://www.unl.edu/chemengr/
University of Nevada, Reno, Chemical and Metallurgical Engineering Department	http://www.scs.unr.edu/unr/
University of New Mexico, Chemical and Nuclear Engineering Department	http://129.24.216.193/
University of New Hampshire, Durham, College of Engineering and Physical Sciences, Chemical Engineering Department	http://unhinfo.unh.edu:70/0/unh/acad/ceps/dept/cheng/index.html
University of New Haven, School of Engineering, Chemistry and Chemical Engineering Department	http://www.newhaven.edu/UNH/Academics/Departments/Chemistry.html#BSChemEng
University of North Dakota, College of Engineering and Mines, Chemical Engineering Department	http://www.und.nodak.edu/sem/chemical.eng/
University of Notre Dame, Department of Chemical Engineering	http://www.nd.edu/~chegdept/

Chemical Engineering Programs *(continued)*

University of Oklahoma, College of Engineering, School of Chemical Engineering and Materials Science	http://www.uoknor.edu/cems/
University of Pennsylvania, Chemical Engineering Department	http://www.seas.upenn.edu/cheme/chehome.html
University of Pittsburgh, School of Engineering, Department of Chemical and Petroleum Engineering	http://www.pitt.edu/~engwww/che.html
University of Puerto Rico, Department of Chemical Engineering	http://www.upr.clu.edu/english/home.html
University of Rhode Island, College of Engineering, Chemical Engineering Department	http://www.egr.uri.edu/chemical.html
University of Rochester, Department of Chemical Engineering	http://www.che.rochester.edu:8080/
University of South Florida, College of Engineering, Chemical Engineering Department	http://www.eng.usf.edu/CE/chemeng.html
University of South Carolina, Department of Chemical Engineering	http://www.engr.scarolina.edu/engr/chem/
University of South Alabama, Chemical Engineering Department	gopher://jaguar1.usouthal.edu/
University of Southern California, School of Engineering, Department of Chemical Engineering	http://www.usc.edu/dept/engr/CHEE/che.html
University of Southwestern Louisiana, Chemical Engineering Department	http://www.usl.edu/

Chemical Engineering Programs *(continued)*

University of Tennessee, Knoxville, Chemical Engineering Department	http://flory.engr.utk.edu/che/che.html
University of Tennessee, Chattanooga, Chemical Engineering Department	gopher://chem.engr.utc.edu:70/1
University of Texas at Austin, Department of Chemical Engineering	http://www.che.utexas.edu/
University of Toledo, College of Engineering, Chemical Engineering Department	http://131.183.20.20/
University of Tulsa, College of Engineering and Applied Sciences, Chemical Engineering Department	http://inti.ce.utulsa.edu/intro.html
University of Utah, College of Engineering, Department of Chemical and Fuels Engineering	http://www.che.utah.edu/
University of Virginia, School of Engineering and Applied Science, Chemical Engineering Department	http://www.cs.virginia.edu/~seas/
University of Washington, College of Engineering, Department of Chemical Engineering	http://www.engr.washington.edu/departments/cheme/
University of Wisconsin, Madison, College of Engineering, Department of Chemical Engineering	http://www.engr.wisc.edu/che/
University of Wyoming, College of Engineering, Department of Chemical and Petroleum Engineering	http://wwweng.uwyo.edu/cheme.html

Chemical Engineering Programs *(continued)*

Vanderbilt University, School of Engineering, Department of Chemical Engineering	http://www.vuse.vanderbilt.edu/~eap/che.htm
Villanova University, College of Engineering, Chemical Engineering Department	http://www.vill.edu/academic/engineer/chem_eng.htm
Virginia Polytechnic Institute and State University, Chemical Engineering Department	http://www.vt.edu:10021/eng/che/index.html
Washington State University, Department of Chemical Engineering	http://www.che.wsu.edu/
Washington University in St. Louis, Chemical Engineering Department	http://wuche.wustl.edu/
Wayne State University, College of Engineering, Chemical Engineering Department	http://www.eng.wayne.edu/CHE.html
West Virginia Institute of Technology, Leonard C. Nelson College of Engineering, Chemical Engineering Department	http://www.wvitcoe.wvnet.edu/Engineering/chemical.html
West Virginia University, College of Engineering, Chemical Engineering Department	http://www.coe.wvu.edu/~wwwche/
Worcester Polytechnic Institute, Chemical Engineering Department	http://www.wpi.edu/Depts/Academic/ChemEng/
Yale University, Faculty of Engineering, Department of Chemical Engineering	http://www.cis.yale.edu/yaleche/
Youngstown State University, Chemical Engineering Department	http://gateway.cis.ysu.edu/

Chemistry Programs

Akron University, Department of Chemistry	http://atlas.chemistry.uakron.edu:8080/
Auburn University, Department of Chemistry	http://www.duc.auburn.edu/chemistry/faculty_interests.html
Bethany College, Chemistry Department	http://info.bethany.wvnet.edu/AboutBethany/Departments/Chemistry/*index.html
Boston College, Merkert Chemistry Center	http://chemserv.bc.edu/
Boston University, Department of Chemistry	http://chem.bu.edu/
Brigham Young University, Department of Chemistry and Biochemistry	http://nmra.byu.edu/welcome.html
Brookhaven National Laboratory, Chemistry Department	http://www.chemistry.bnl.gov/chemistry.html
Brown University, Department of Chemistry	http://www.chem.brown.edu/index.html
Butler University, Department of Chemistry	http://www.butler.edu/www/chemistry
California State University Stanislaus, Chemistry Department	http://wwwchem.csustan.edu/
California State University, Fresno, Chemistry Department	http://129.8.104.30:8080/
Caltech, Division of Chemistry and Chemical Engineering	http://www.caltech.edu/caltech/Chemistry.html
Calvin College, Department of Chemistry and Biochemistry	http://www.calvin.edu/~grayt/Chemistry_at_Calvin.html

Chemistry Programs *(continued)*

Carnegie Mellon University, Department of Chemistry	http://www.chem.cmu.edu/
Case Western Reserve University, Department of Chemistry	http://chemwww.cwru.edu/
Case Western Reserve University, Biochemistry Department	http://biochemistry.bioc.cwru.edu/
Central Instrument Facility	http://www1.chm.colostate.edu/
Centre College, Chemistry Department	http://www.centre.edu/~miles/che/index.html
Chemical Industry Institute of Toxicology	http://www.ciit.org/HOMEP/ciit.html
Chemical Physics Preprint Database	http://www.chem.brown.edu/chem-ph.html
Colorado State University, Chemistry Department	http://www.colostate.edu/Depts/Chem/ index.html
Columbia University, Department of Chemistry	http://www.cc.columbia.edu/~chempub/
Connecticut College, Chemistry Department	http://camel.conncoll.edu/ccacad/zimmer/mz2.html
Cooper Union, Department of Chemistry	http://www.cooper.edu/engineering/chemechem/depts_info/chemdept.html
Cornell University, Department of Chemistry	http://www.chem.cornell.edu/
CUNY, Hunter College, Chemistry Department	http://patsy.hunter.cuny.edu:8001/chemistry.html
Duke University, Department of Chemistry	http://www.chem.duke.edu/

Chemistry Programs *(continued)*

East Tennessee State University, Department of Chemistry	http://www.east-tenn-st.edu/~chemists/
Emory University, Chemistry Department	http://www.emory.edu/CHEMISTRY/
Emory University, The Cherry L. Emerson Center for Scientific Computation at the Chemistry Department	http://www.chem.emory.edu
Florida State University, Chemistry Department	http://www.chem.fsu.edu/
George Washington University, Department of Chemistry	http://www.gwu.edu/~gwchem/
Georgetown University, Department of Chemistry	http://www.georgetown.edu/departments/chemistry/chemistry.html
Georgia Institute of Technology, School of Chemistry and Biochemistry	http://www.chemistry.gatech.edu/
Hahnemann University, Department of Biochemistry	http://www.hahnemann.edu/heme-iron/netwelcome.html
Hamline University, Chemistry Department	http://www.hamline.edu/depts/chemistry/index.html
Harvard University, Department of Chemistry	http://www-chem.harvard.edu/
Illinois State University, Department of Chemistry	http://neon.che.ilstu.edu/
Indiana University Purdue University Indianapolis (IUPUI), Department of Chemistry	http://chem.iupui.edu/homepage.html
Iowa General Chemistry Network: A FIPSE Project	http://www.public.iastate.edu/~fipse-chem/homepage.html

Chemistry Programs *(continued)*

Iowa State University, Chemistry Department	http://www.public.iastate.edu/~chemistry/homepage.html
Lebanon Valley College, Chemistry Department	http://www.lvc.edu/www/chemistry/index.html
Lehigh University, Department of Chemistry	http://www.lehigh.edu/~inche/inche.html
Loras College Department of Chemistry, Iowa	http//192.152.29.158/
Los Alamos National Laboratory, Chemical Science and Technology Division	http://mwanal.lanl.gov/
Louisiana State University, Department of Chemistry	http://chrs1.chem.lsu.edu/
Loyola Marymount University, Department of Chemistry and Biochemistry	http://chem.lmu.edu/chem/home.html
Loyola University Chicago, Department of Chemistry	http://www.luc.edu/depts/chem
M.I.T., Department of Chemistry	http://web.mit.edu/afs/athena.mit.edu/org/c/chemistry/www/chem-home.html
Marshall University, Department of Chemistry	http://www.marshall.edu/chemistry/
Michigan State University, Chemistry Department	http://www.cem.msu.edu/
Michigan Technological University, Division of Chemical Sciences	http://www.chem.mtu.edu/
Middle Tennessee State University, Department of Chemistry, Amazing Science at the Roxy (Physical Science Education)	http://161.45.2.50

Chemistry Programs *(continued)*

Mississippi State University, Department of Chemistry	http://www.MsState.edu/Dept/Chemistry/
Missouri Western State College, Chemistry	http://www.mwsc.edu/~chemist/
National Center for Supercomputing Applications, NCSA Digital Information System, Chemistry	http://www.ncsa.uiuc.edu/SCMS/DigLib/text/chemistry/Chemistry.html
National Center for Supercomputing Applications, NCSA ChemViz Group, features materials for high school students and teachers	http://eads.ncsa.uiuc.edu/~datkins/
National Institutes of Health, Molecular Recognition Section, provides mini-encyclopedia of purine research	http://mgddk1.niddk.nih.gov:8000
National Institute of Standards and Technology (NIST)	http://www.nist.gov/
National Institutes of Health, Molecular Modeling Home Page	http://www.nih.gov/molecular_modeling/mmhome.html
New York University Medical Center, Protein Chemistry	http://128.122.10.3/index.htm
North Carolina State University, Department of Chemistry	http://www2.ncsu.edu/ncsu/chemistry/chem.html
Northern Illinois University, Department of Chemistry	http://hackberry.chem.niu.edu:70/0/webpage.html
Northern Illinois University, Chemists Address/Phone Book and directions on how to add your name to this index	http://hackberry.chem.niu.edu:70/0/ChemDir/index.html
Oberlin College, Department of Chemistry	http://chemserver.chem.oberlin.edu/

Chemistry Programs *(continued)*

Ohio State University, Department of Chemistry	http://www.chemistry.ohio-state.edu/
Ohio Supercomputer Center, Computational Chemistry List	http://www.osc.edu/chemistry.html
Ohio University, Department of Chemistry	http://quanta.phy.ohiou.edu/
Oklahoma State University, Department of Chemistry	http://bubba.ucc.okstate.edu/jgelder/Main.html
Oregon Graduate Institute of Science & Technology, Department of Environmental Science and Engineering	http://www.ese.ogi.edu/
Oregon Graduate Institute of Science & Technology, Department of Chemistry, Biochemistry and Molecular Biology	http://amethyst.cbs.ogi.edu/
Oregon State University, Department of Chemistry	http://www.chem.orst.edu/
Pacific Lutheran University, Department of Chemistry	http://www.plu.edu/www/chem/chemdept.html
Pacific Northwest Laboratories, Gaussian Basis Set Order Form	http://www.emsl.pnl.gov:2080/forms/basisform.html
Pacific Northwest Laboratories, Environmental and Molecular Sciences Laboratory	http://www.emsl.pnl.gov:2080/
Pennsylvania State University, Department of Chemistry	http://www.chem.psu.edu/
Pomona College, Chemistry Department	http://patricia.pomona.edu/
Princeton University, Department of Chemistry	http://www.princeton.edu/~chemdept/

Chemistry Programs *(continued)*

Purdue University, Chemistry Department	http://www.chem.purdue.edu/
Purdue University, Department of Medicinal Chemistry Pharmacognosy	http://www.purdue.edu/Pharmacy/mdchdept.html
Purdue University, Food Science Department	http://www.foodsci.purdue.edu/
Rensselaer Polytechnic Institute, Department of Chemistry	http://www.rpi.edu/dept/chem/rpichem/chemhome.html
Rice University, Department of Chemistry	http://pchem1.rice.edu/RiceChem.html
Rockefeller University, Laboratory of Physical Biochemistry	http://mriris.rockefeller.edu/
Rutgers University, Department of Chemistry	http://chmwww.rutgers.edu/
Rutgers University, Nanotechnology Papers	http://planchet.rutgers.edu/
Sam Houston State University, Department of Chemistry	http://www.shsu.edu/~chemistry/
Scripps Research Institute, Graduate Program in Chemistry	http://www.scripps.edu/pub/chmgrads/chemhome.html
Southern Illinois University at Carbondale, Department of Chemistry and Biochemistry	http://www.science.siu.edu/chemistry/
Stanford University, Chemistry Department	http://www-chem.stanford.edu/
Stanford University, Journal of Biological Chemistry	http://www-jbc.stanford.edu/jbc/
State University of New York at Albany, Department of Chemistry	http://www.albany.edu/~bjl01/

Chemistry Programs *(continued)*

State University of New York at Binghamton, Department of Chemistry	http://chemiris.chem.binghamton.edu:8080/
State University of New York at Stony Brook (USB), Bioinorganic Chemistry Server	http://sbchm1.sunysb.edu/koch/biic.html
State University of New York at Stony Brook (USB), Department of Chemistry	http://sbchm1.sunysb.edu/
State University of New York at Buffalo, Department of Chemistry	http://www.chem.buffalo.edu/
Texas A&M University, Laboratory for Magnetic Resonance and NMR Periodic Table	http://hawserv80.tamu.edu/hawhomepage/Haw1.html
Texas A&M University, Chemistry Department	http://wwwchem.tamu.edu/
Texas Christian University (TCU), Chemistry Department	http://www.chm.tcu.edu/www/chm/tcuchem1.html
Texas Tech University, Chemistry and Biochemistry Department	http://molec1.chem.ttu.edu/
Tufts University, Department of Chemistry	http://www.tufts.edu/departments/chemistry/
University of Georgia, Department of Chemistry, Center for Computational Quantum Chemistry	http://zopyros.ccqc.uga.edu/
University of Notre Dame, Mass Spectrometry Laboratories	http://www.nd.edu/~dgriffit/MassSpec/masspec.html
University of Notre Dame, Department of Chemistry	http://www.science.nd.edu/chemistry/chemistry.html
University of Alabama, Department of Chemistry	http://www.as.ua.edu/chemistry/

Chemistry Programs *(continued)*

University of Alabama at Birmingham, Department of Chemistry	http://sleeper.chem.uab.edu/
University of Alabama in Huntsville, Chemistry Department	http://matsci.uah.edu/Chemistry/
University of California Berkeley, College of Chemistry	http://www.cchem.berkeley.edu/index.html
University of California Irvine, Department of Chemistry	http://www.chem.uci.edu/
University of California Los Angeles, Department of Chemistry and Biochemistry	http://www.chem.ucla.edu/dept/Chemistry.html
University of California, Santa Cruz, Chemistry Department	http://www.chemistry.ucsc.edu/
University of California Davis, Department of Chemistry	http://www-chem.ucdavis.edu/
University of California San Diego, Department of Chemistry and Biochemistry	http://checfs1.ucsd.edu/
University of California Santa Barbara, Department of Chemistry	http://128.111.114.72/
University of California San Francisco, Department of Pharmaceutical Chemistry	http://www.pharm.ucsf.edu/
University of California Riverside, Department of Chemistry	http://www.chem.ucr.edu/
University of Chicago, Department of Chemistry	http://rainbow.uchicago.edu/chemistry/
University of Cincinnati, Department of Chemistry	http://www.che.uc.edu/
University of Connecticut, Storrs, Chemistry Department	http://spirit.lib.uconn.edu/Chemistry/

Chemistry Programs *(continued)*

University of Delaware, Department of Chemistry and Biochemistry	http://www.udel.edu/chem/chem.html
University of Denver, Chemistry Department	http://www.du.edu/~jgilbert/chemhome.html
University of Florida, Department of Chemistry	http://www.chem.ufl.edu/
University of Florida, Drago Group	http://inorganic1.chem.ufl.edu/
University of Hawaii, Department of Chemistry	http://krypton.nmr.hawaii.edu/UH_Chem
University of Houston, Departmen of Chemistry	http://www.chem.uh.edu/
University of Idaho, Department of Chemistry	http://www.chem.uidaho.edu/
University of Illinois at Chicago, Chemistry Department	http://gopher.chem.uic.edu/
University of Illinois at Urbana-Champaign, Chemistry Department	http://www.scs.uiuc.edu/
University of Kentucky, Department of Chemistry	http://www.uky.edu/ArtsSciences/Chemistry/
University of Maryland Baltimore County (UMBC), Department of Chemistry and Biochemistry	http://www.research.umbc.edu/~smith/chem/chem.html
University of Maryland College Park, Department of Chemistry and Biochemistry	http://www-chem.umd.edu/
University of Massachusetts, Department of Chemistry	http://128.119.52.169/

Chemistry Programs *(continued)*

University of Massachusetts Lowell, Chemistry Department	http://woods.uml.edu/www/Chemistry_Dept/chemistry.html
University of Memphis, Department of Chemistry	http://www.chem.memphis.edu/umchem.html
University of Miami, Department of Chemistry	http://www.ir.miami.edu/CHM/
University of Minnesota, Department of Chemistry	http://www.chem.umn.edu/
University of Mississippi, Department of Chemistry	http://www.olemiss.edu/depts/chemistry/
University of Missouri-St. Louis, Department of Chemistry, Chemistry Textbook Archive	http://www.umsl.edu/divisions/artscience/chemistry/books/welcome.html
University of Missouri-Rolla, Chemistry Department	http://www.chem.umr.edu/
University of Missouri-St. Louis, Department of Chemistry	http://www.umsl.edu/divisions/artscience/chemistry/chemistry.html
University of Nebraska at Kearney, Department of Chemistry	http://www.unk.edu/departments/chemistry
University of Nebraska-Lincoln, Department of Chemisry	http://wwitch.unl.edu/unl_chem.html
University of Nevada, Reno, Department of Chemistry	http://www.chem.unr.edu/
University of New Mexico, Department of Chemistry	http://www.unm.edu/~dmclaugh/home.html
University of New Orleans, Department of Chemistry	http://www.chem.uno.edu/

Chemistry Programs *(continued)*

University of New Mexico, Chemistry Principles (Textbook, requires a PostScript browser)	http://www.unm.edu/~dmclaugh/Principles.html
University of New Mexico, Enke Research Laboratory, Mass Spectrometry	http://enke.unm.edu/
University of North Carolina at Chapel Hill, Department of Chemistry	http://net.chem.unc.edu/
University of Oklahoma, Department of Chemistry and Biochemistry	http://cheminfo.chem.uoknor.edu/
University of Oregon, Chemistry Department	http://www-vms.uoregon.edu/~chem/index.html
University of Pennsylvania, Department of Chemistry	http://www.chem.upenn.edu/
University of Pittsburgh, Department of Chemistry	http://www.pitt.edu/~chemrdc/chemistry.html
University of Rhode Island, Department of Chemistry	http://www.chm.uri.edu/index.html
University of Rochester, Department of Chemistry	http://rhs.chem.rochester.edu/
University of South Florida, Department of Chemistry	http://nosferatu.cas.usf.edu/chemistry/index.html
University of South Dakota, Department of Chemistry	http://www.usd.edu/chemistry
University of Southern California, Chemistry Department	http://www.usc.edu/dept/chemistry/home.html
University of Southern Mississippi, Polymer Science Research Center	http://www.usm.edu/psrc/dps-menu.html

Chemistry Programs *(continued)*

University of Southern Mississippi, Department of Chemistry and Biochemistry	http://puddle.st.usm.edu/Chem.html
University of Southern Colorado, Chemistry Department	http://meteor.uscolo.edu/chem/
University of Tennessee, Knoxville, Department of Chemistry	http://novell.chem.utk.edu/
University of Texas, Medical Branch in Galveston, Department of Human Biological Chemistry and Genetics	http://www.hbcg.utmb.edu/
University of Texas at Arlington, Department of Chemistry and Biochemistry	http://utachem.uta.edu/
University of Texas, Medical Branch in Galveston, Internet NMR Resources	http://www.nmr.utmb.edu/nmrlist.html
University of Texas, Medical Branch in Galveston, University of Texas Medical Branch, NMR Center	http://www.nmr.utmb.edu/
University of Texas at Austin, Chemistry Library	http://www.lib.utexas.edu/Libs/Chem/first/mallet.html
University of Texas at Austin, Department of Chemistry and Biochemistry	http://www.cm.utexas.edu/
University of Utah, Interdepartmental Graduate Program in Biological Chemistry	http://lysine.pharm.utah.edu/biolchem/Bchome.html
University of Utah, Department of Chemistry	http://www.chem.utah.edu/
University of Washington, Department of Chemistry	http://www.chem.washington.edu/

Chemistry Programs *(continued)*

University of Wisconsin-Madison, Department of Chemistry, and New Traditions Curriculum Project	http://www.chem.wisc.edu/
University of Wisconsin-Oshkosh, Chemistry Department	http://www.uwosh.edu/home_pages/departments/chemistry/home.html
University of Wisconsin, School of Pharmacy, Molecular Modeling Laboratory	http://www.pharmacy.wisc.edu/
University of Wyoming, Chemistry Department	http://www.uwyo.edu/a&s/chem/chemistry.htm
Villanova University, Department of Chemistry	http://rs6chem.vill.edu/
Virginia Tech, Department of Chemistry, Chemistry Hypermedia Project, and Ethics in Science	http://www.chem.vt.edu/
Washington University in St. Louis, Department of Chemistry and Center for Molecular Design	http://wunmr.wustl.edu
West Virginia University, Department of Chemistry	http://www.as.wvu.edu/chemistry/
Western Michigan University, Chemistry Department	http://www.wmich.edu/chemistry/
Widener University, Department of Chemistry	http://science.widener.edu/chemistry.html
Wilkes University, Department of Chemistry	http://www.wilkes.edu/WilkesDocs/SSEHome.html
Worcester Polytechnic Institute, Department of Chemistry	http://www.wpi.edu/Depts/Academic/Chemistry/

Yale University, Department of Chemistry	http://www.chem.yale.edu/

Environmental Engineering Programs

ARL Home Page - Penn State University	http://www.arl.psu.edu/index.html
Brigham Young University, School of Engineering	http://www.et.byu.edu/centers/ce-info.html
Carnegie Mellon University, Dept. of Civil and Environmental Engineering	http://www.ce.cmu.edu:8000/
Center for Clean Technology at University of California, Los Angeles	http://cct.seas.ucla.edu
Department of Environmental Science and Engineering, Oregon Graduate Institute of Science and Technology	http://www.ese.ogi.edu/
University of Cincinnati, Environmental Engineering and Science	http://www.cee.uc.edu:80/~eewww/
Georgia Tech, School of Civil and Environmental Engineering	http://howe.ce.gatech.edu/ce.html
New Mexico Institute of Mining and Technology, Environmental Engineering Department	http://www.nmt.edu:80/nmt/dept/enveng/
Stanford University, Civil Engineering	http://www-ee.stanford.edu/cive.html
Texas A&M University Department of Civil Engineering	http://info-civil.tamu.edu/

Environmental Engineering Programs *(continued)*

Thayer School of Engineering	http://caligari.dartmouth.edu/thayer/thayer.html
University of Florida Civil Engineering, General Information	http://www.ce.ufl.edu/brochure/brochure.html

CHAPTER 4

DISCUSSION LISTS

Chapter 4

Discussion Lists

AspenTech Info E-Mail Discussion Lists

Description: Aspen-use is an unmoderated e-mail list used to discuss the topics such as general information, tips, techniques, applications, experiences, sources for examples, porting models from other systems, and related software.

Ati-vl is also an unmoderated e-mail discussion list that has been established as a forum for rapid exchange of information, ideas, and opinions related to the management and development of the AspenTech Info virtual libraries.

It is expected that these lists will allow users and developers to assist each other as well as distribute hints and other information about AspenTech Info-related projects.

Subscription: listproc@listproc.che.ufl.edu

Source: http://www.che.ufl.edu/~atinfo/discuss/lists.index.html#aspen-use

CHEMCOM

Description: Chemistry in the community discussion list.

Subscription: listserv@ubvm.cc.buffalo.edu

Source: http://www.tile.net/tile/listserv/chemcom.html

CHEMCONF

Description: Conferences on chemistry research and education.

Subscription: send message to "sub CHEMCONF your name to listserv%umdd.bitnet@listserv.net"

Source: http://www.tile.net/tile/listserv/chemconf.html

CHEMCORD

Description: A forum for coordinators of general chemistry programs and labs to discuss issues important to members.

Subscription: listserv@umdd.umd.edu

Source: http://www.n2n2.com/KOVACS/CD/0441.html

CHEME-L

Description: Chemical engineering mailing list. It addresses the interests and concerns of the chemical engineering and science community. The Cheme-L is noncommercial and maintained ad hoc, and does not endorse any organization or entity.

Subscription: cheme-l@ulkyvm.louisville.edu

Sources: cheme-l@ulkyvm.louisville.edu/~r0mira01/cheme-l

CHEMED-L

Description: Discussion of chemistry education.

Subscription: listserv@uwf.cc.uwf.edu

Source: http://www.n2n2.com/KOVACS/CD/0443.html

Chemical Bulletin Board System

Description: Information provided includes the latest information on chemical products, update of materials safety data sheets, chemical database system, on-line shopping mall, and user-friendly graphical interface.

Subscription: lehan@VBS.Com

Source: http://www.cstone.net/~vine/cbbs/cbbs.html

Chemical Engineering Newsletter

Description: The Chemical Engineering Newsletter is available to anyone with E-mail access.

Subscription: listproc@cae.wisc.edu

Sources: http://www.che.ufl.edu/WWW-CHE/topics/newsgroups.html

CHEMINF-L

Description: Discussion of chemical information sources related to chemistry.

Subscription: listserv@iubvm.indiana.edu

Source: http://www.n2n2.com/KOVACS/CD/0446.html

CHEMISTRY

Description: Discussion of computational chemistry and others.

Subscription: chemistry-request@ccl.osc.edu

Source: http://www.n2n2.com/KOVACS/CD/0444.html

CHEMLAB-L

Description: Discussion of chemistry laboratories, students' experiments, classroom demonstrations and shows for the public of chemical processes, chemistry stockroom management, lab safety, and small-scale chemical waste handling procedures.

Subscription: listserv@Beaver.Bemidji.MSUS.edu

Source: http://www.n2n2.com/KOVACS/CD/0445.html

CICOURSE

Description: A forum in which people can teach and learn about chemical information sources through workshops, courses, etc.

Subscription: listserv@iubvm.indiana.edu

Source: http://www.n2h2.com/KOVACS/CD/0447.html

ECDM Internet Mailing List

Description: Discussion of environmentally conscious design and manufacturing.

Subscription: listserv@uwindsor.ca

Source: http://ie.uwindsor.ca/ecdmlist/welcome.html

EPA - Air and Radiation

Contact: internet_support@unixmail.rtpnc.epa.gov

Source: http://www.epa.gov/epahome/Offices.html

EPA-Administration and Resources Management

Contact: internet_support@unixmail.rtpnc.epa.gov

Source: http://www.epa.gov/epahome/Offices.html

EPA-National Air and Radiation Environmental Laboratory (NAREL)

Contact: internet_support@unixmail.rtpnc.epa.gov

Source: http://www.epa.gov/epahome/Labs.html

EPA-National Exposure Research Lab (NERL) - Athens GA

Contact: internet_support@unixmail.rtpnc.epa.gov

Source: http://www.epa.gov/epahome/Labs.html

EPA-National Exposure Research Lab (NERL)

Contact: internet_support@unixmail.rtpnc.epa.gov

Source: http://www.epa.gov/epahome/Labs.html

EPA-National Health & Environmental Effects Research Lab (NHEERL)

Contact: internet_support@unixmail.rtpnc.epa.gov

Source: http://www.epa.gov/epahome/Labs.html

EPA-National Risk Management Research Lab (NRMRL)

Contact: internet_support@unixmail.rtpnc.epa.gov

Source: http://www.epa.gov/epahome/Labs.html

EPA-National Vehicle and Fuel Emissions Laboratory

Contact: internet_support@unixmail.rtpnc.epa.gov

Source: http://www.epa.gov/epahome/Labs.html

EPA-Office of Radiation & Indoor Air-Las Vegas Laboratory

Contact: internet_support@unixmail.rtpnc.epa.gov

Source: http://www.epa.gov/epahome/Labs.html

EPA-Policy Planning and Evaluation

Contact: internet_support@unixmail.rtpnc.epa.gov

Source: http://www.epa.gov/epahome/Offices.html

EPA-Radiation Surveillance Laboratory

Contact: internet_support@unixmail.rtpnc.epa.gov

Source: http://www.epa.gov/epahome/Labs.html

EPA-Region 1 (ME, NH, VT, MA, RI, CT)

Contact: internet_support@unixmail.rtpnc.epa.gov

Source: http://www.epa.gov/epahome/Regions.html

EPA-Region 2 (NY, NJ, PR, VI)

Contact: internet_support@unixmail.rtpnc.epa.gov

Source: http://www.epa.gov/epahome/Regions.html

EPA-Region 3 (PA, DE, DC, MD, VA, WV)

Contact: internet_support@unixmail.rtpnc.epa.gov

Source: http://www.epa.gov/epahome/Regions.html

EPA-Region 4 (KY, TN, NC, SC, MS, AL, GA, FL)

Contact: internet_support@unixmail.rtpnc.epa.gov

Source: http://www.epa.gov/epahome/Regions.html

EPA-Region 5 (MN, WI, IL, MI, IN, OH)

Contact: internet_support@unixmail.rtpnc.epa.gov

Source: http://www.epa.gov/epahome/Regions.html

EPA-Region 6 (NM, TX, OK, AR, LA)

Contact: internet_support@unixmail.rtpnc.epa.gov

Source: http://www.epa.gov/epahome/Regions.html

EPA-Region 7 (NE, KS, IA, MO)

Contact: internet_support@unixmail.rtpnc.epa.gov

Source: http://www.epa.gov/epahome/Regions.html

EPA-Region 8 (MT, ND, WY, SD, UT, CO)

Contact: internet_support@unixmail.rtpnc.epa.gov

Source: http://www.epa.gov/epahome/Regions.html

EPA-Region 9 (CA, NV, AZ, HI)

Contact: internet_support@unixmail.rtpnc.epa.gov

Source: http://www.epa.gov/epahome/Regions.html

EPA-Region 10 (WA, OR, ID, AK)

Contact: internet_support@unixmail.rtpnc.epa.gov

Source: http://www.epa.gov/epahome/Regions.html

EPA-Research and Development

Contact: internet_support@unixmail.rtpnc.epa.gov

Source: http://www.epa.gov/epahome/Offices.html

EPA-Robert S. Kerr Environmental Research Lab - Ada OK

Contact: internet_support@unixmail.rtpnc.epa.gov

Source: http://www.epa.gov/epahome/Labs.html

EPA-Solid Waste and Emergency Response

Contact: internet_support@unixmail.rtpnc.epa.gov

Source: http://www.epa.gov/epahome/Offices.html

EPA-Water

Contact: internet_support@unixmail.rtpnc.epa.gov

Source: http://www.epa.gov/epahome/Offices.html

GEOSYN

Description: A forum for communications among engineers, contractors, manufacturers involved with geosynthetics.

Subscription: majordomo@csn.org

Source: http://www.n2h2.com/KOVACS/CD/0449.html

HIRIS-L

Description: High resolution infrared spectroscopy list. This is a forum for discussion on infrared spectroscopy instruments.

Subscription: listserv@icineca.cineca.it

Source: http://www.n2h2.com/KOVACS/CD/0450.html

ICS-L

Description: International Chemometrics Society list. ICS is concerned with the development and application of mathematical and statistical methods for the analysis of chemical data.

Subscription: listserv@umdd.umd.edu

Source: http://www.n2h2.com/KOVACS/CD/0451.html

Info-labVIEW mailing list

Description: The list is used by LabVIEW users to communicate with other users and with National Instruments.

Subscription: info-labview-request@pica.army.mil

Source: http://www.che.ufl.edu/WWW-CHE/topics/newsgroups.html

Introduction to the SAFETY mailing list

Description: SAFETY is an Internet electronic mailing list. It discusses environmental and occupational health and safety issues, particularly those associated with college and university campuses, although a wide range of subjects is encouraged.

Subscription: gopher://siri.uvm.edu/00ftp%3aSIRI%3aSAFETY_list_information%3aSAFETY_WELCOME

Source: http://www.che.ufl.edu/WWW-CHE/topics/newsgroups.html

MAT-DSGN

Description: Forum on materials designed by computer.

Subscription: listserv@jpntuvm0.bitnet

Source: http://www.n2h2.com/KOVACS/CD/0455.html

PAD

Description: Polymer analysis and characterization discussions.

Subscription: PAD-Request@listserv.syr.edu

Source: http://www.n2h2.com/KOVACS/CD/0461.html

Surface-l

Description:	For discussion on surface science, techniques, instrumentation, analytical procedure, data interpretation and possible applications.
Subscription:	surface-request@surf.ssw.uwo.ca
Source:	http://www.n2h2.com/KOVACS/CD/0468.html

CHAPTER 5

NEWSGROUPS

Chapter 5

Newsgroups

---

sci.answers

Description: For the information exchange related to any scientific subjects.

Sample subjects: Conventional fusion FAQ, organ transplantation, and salt water.

---

sci.chem.organomet

Description: For the information exchange related to organometallic compounds.

Sample subjects: Catalysts and monomethylarsonic acid (MMA).

---

sci.chem

Description: For the information exchange related to chemistry and chemical engineering.

Sample subjects: Immobilization of carbon dioxide, super critical water oxidation, and teflon and sublimation.

---

sci.cryonics

Description: For the information exchange related to cryogenics.

Sample subjects: Cryonic FAQ (frequently asked questions), and cryonic suspension, which is an experimental procedure whereby patients who can no longer be kept alive with today's medical abilities are preserved at low temperatures for treatment in the future.

sci.data.formats

Description: For the information exchange related to data management subjects.

Sample subjects: Video comparisons and multimedia information.

sci.edu

Description: For the information exchange related to education.

Sample subjects: Science fair, questions on grading, and physics.

sci.electronics

Description: For the information exchange related to electronics.

Sample subjects: Microcontroller FAQ, embedded processor, and equivalent transistor.

sci.energy.hydrogen

Description: For the information exchange related to hydrogen power.

Sample subjects: Fuel cell, storage of hydrogen, and liquid hydrogen.

sci.energy

Description: For the information exchange related to energy subjects.

Sample subjects: Nuclear power, global warming, and electricity.

sci.engr.biomed

Description: For the information exchange related to biomedical engineering.

Sample subjects: Measurement of oxygen content in blood, membrane, and surplus equipment.

sci.engr.chem

Description: This group is similar to the group, "news:sci.chem," which is to provide the information exchange related to chemistry and chemical engineering.

Sample subjects: Doxone content of 2,4-D, how to melt magnesium, and sulphuric acid.

sci.engr

Description: For the information exchange related to engineering.

Sample subjects: Engineering education, plutonium, and quality and engineering.

sci.environment

Description: For the information exchange related to environmental aspects.

Sample subjects: Environmental protection, ISO 1400, and climate change.

sci.fractals

Description: For the information exchange related to fractals.

Sample subjects: Discrete dynamics, fractal math, and fractal engine.

---

sci.geo.fluids

Description: For the information exchange related to geological fluid phenomena.

Sample subjects: Lava flow, ocean modeling, and geoscience.

---

sci.image.processing

Description: For the information exchange related to image processing.

Sample Subjects: Digital medical images, image resolution, and motion measurement.

---

sci.materials

Description: For the information exchange related to materials science.

Sample subjects: Corrosion, vacuum equipment, and color changing materials.

---

sci.math

Description: For the information exchange related to mathematics.

Sample subjects: Mapping function, multidimensional space, and logarithms.

---

sci.mech.fluids

Description: For the information exchange related to mechanical fluids.

Sample subjects: Wave speed, density, and viscosity.

sci.misc:

Description: For the information exchange related to any scientific miscellanies.

Sample subjects: Pyramids, air measurement, and trips.

sci.nanotech

Description: Nano means one billionth (10^{-9}) part of. This group is for the information exchange related to nano-technologies.

Sample subjects: Infiltration of nano-technology into everyday life.

sci.nonlinear

Description: For the information exchange related to nonlinear systems.

Sample subjects: Fluidized beds and chaos, chaotic behavior in signal transduction, and chaos definitions.

sci.op-research

Description: For the information exchange related to operational research.

Sample subjects: Hundred percent rule, human-aided optimization, and multispectral images.

sci.optics

Description: For the information exchange related to optics.

Sample subjects: Acoustic holography, absorptivity of limestone, and crystran optical crystals.

sci.physics.accelerators

Description: For the information exchange related to accelerators.

Sample subjects: Particle, electrons, and mass.

sci.physics.fusion

Description: For the information exchange related to fusion phenomena.

Sample subjects: Fusion, fusion glossary, and heat.

sci.physics.particle

Description: For the information exchange related to particle physics.

Sample subjects: Radiation, future of physics, and gravity.

sci.physics.research

Description: For the information exchange related to research in physics.

Sample subjects: Noise, multi-dimension, and variation.

sci.physics

Description: For the information exchange related to physics.

Sample subjects: Immortality of physics, light, and sound.

sci.polymers

Description: For the information exchange related to polymers and plastics.

Sample subjects: Used equipment, ceramics, and impermeable membrane.

---

sci.research.careers

Description: For the information exchange related to research in career development.

Sample subjects: Faculty openings; electromagnetic jobs; and business.

---

sci.research

Description: For the information exchange related to research in science.

Sample subjects: Air measurement, scan application, and human superpower.

---

sci.space.science

Description: For the information exchange related to space science.

Sample subjects: Cosmology, expansion of the universe, and orbits.

---

sci.stat.math

Description: For the information exchange related to statistical mathematics.

Sample subjects: Experimental design, fussy logic, and statistics.

---

sci.systems

Description: For the information exchange related to scientific systems.

Sample subjects: System thinking, web site, and autopoiesis.

sci.techniques.xtallography

Description: For the information exchange related to xtallographies.

Sample subjects: Simulation of x-ray diffraction, crystal alignment, and nitrogen triiodide.

sci.techniques.microscopy

Description: For the information exchange related to microscopies.

Sample subjects: Dissolving silicon, online microscopy, gold particle counting using NIH image.

sci.techniques.mag-resonance

Description: For the information exchange related to resonance phenomena.

Sample subjects: Used equipment, impact users, and help with inverse and paramagnetic relaxation agents.

sci.techniques.mass-spec

Description: For the information exchange related to mass spectrometry.

Sample subjects: Ion pairs and software.

sci.virtual-worlds

Description: For the information exchange related to scientific virtual worlds.

Sample subjects: Stereoscopic display, alternative physics, and geometries.

CHAPTER 6

GOPHER RESOURCES

Chapter 6

Gopher Resources

American Chemical Society Publications	gopher://infx.infor.com:4500
American Chemical Society Gopher	gopher://acsinfo.acs.org
Australian Defense Academy, Chemistry Listings	gopher://ccadfa.cc.adfa.oz.au/77/ts?chemistry
Center for Disease Control (CDC and Prevention)	gopher://gopher.cdc.gov
Cornell Theory Center, Computational Chemistry Newsletter	gopher://blanca.tc.cornell.edu/11/Forefronts/Computational.Chemistry.News
CSC Chemistry Gopher in Finland	gopher://gopher.csc.fi/11/tiede
Duke University, Department of Chemistry	gopher://gopher.chem.duke.edu/
Environment and Ecology	gopher://riceinfo.rice.edu/11/Subject/Environment
Imperial College, Department of Chemistry	gopher://gopher.ch.ic.ac.uk/
Indiana University, Chemistry Library Gopher	gopher://libgopher.lib.indiana.edu:7050/11/chemlibrary
Johns Hopkins University, Computational Biology	gopher://gopher.gdb.org/
Journal of Chemical Physics Express Service	gopher://jcp.uchicago.edu/
Library of Congress (LOC)	gopher://gopher.loc.gov
Macquarie University, School of Chemistry Gopher	gopher://gopher.chem.mq.edu.au:70/
Material Safety Data Sheets (MSDS)	gopher://atlas.chem.utah.edu/11/MSDS

Michigan State University, Chemistry Gopher	gopher://gopher.cem.msu.edu/1
National Institutes of Health (NIH)	gopher://gopher.nih.gov
National Institute of Environmental Health (NIEHS)	gopher://gopher.niehs.nih.gov
National Library of Medicine (NLM)	gopher://gopher.nlm.nih.gov
National Oceanic and Atmospheric Administration (NOAA) Environmental Information Services	gopher://gopher.esdim.noaa.gov
National Oceanic and Atmospheric Administration (NOAA) Environmental Information Services	gopher://gopher.noaa.gov
North Carolina State University, Chemistry Software	gopher://dewey.lib.ncsu.edu/11/library/disciplines/chemistry/software
Northern Illinois University, Department of Chemistry	gopher://hackberry.chem.niu.edu/1
Northwestern University, Chemistry Gopher	gopher://nuinfo.nwu.edu/00/schools/school/cas/introbulletins/Chemistry
Nucleic Acid Database (NDB) Archive	gopher://ndbserver.rutgers.edu:70/11/ etc/ndb_link_files
Pollution Prevention Program Database	gopher://gopher.pnl.gov:2070/1/.pprc
QCPE Server	gopher://gopher.gdb.org/1ftp%3aqcpe6.chem.indiana.edu%40/
Regensburg Theoretical Chemistry Gopher Server	gopher://rchs1.chemie.uni-regensburg.de/11/
Rice University, Chemistry Information	gopher://chico.rice.edu/11/Subject/Chemistry

Sewage in Puget Sound	gopher://futureinfo.com/1/menu3/menu5/
Titlenet	gopher://gopher.infor.com/1
UNICAMP, IQ Resources	gopher://gopher.iqm.unicamp.br/
University of Tennessee, Department of Chemistry	gopher://neon.chem.utk.edu/1
University of Utah, Department of Chemistry	gopher://atlas.chem.utah.edu/1
University of Wisconsin-Madison, Department of Chemistry	gopher://gopher.chem.wisc.edu/1
University of California Davis, Department of Chemistry	gopher://gopherchem.ucdavis.edu/11/Index/ChemSites_ac
University of California Santa Barbara, Chemistry Listings	gopher://ucsbuxa.ucsb.edu:3001/11/.Sciences/.Chemistry
University of Houston, Analytical Chemistry Center	gopher://oac.hsc.uth.tmc.edu/11/hsc_info/Support%20Services/anal_chem
University of Illinois, Chemistry Gopher	gopher://gopher.uiuc.edu/11/UI/Dinfo/chemsch
University of Missouri-St. Louis, Chemistry Gopher	gopher://slvaxa.umsl.edu/11gopher_root_chem%3a%5b00000\%5d
VERONICA	gopher://veronica.scs.unr.edu/11/veronica
Washington and Lee University, Chemistry info via Gopher	gopher://liberty.uc.wlu.edu:1020/1chemistry%20-sa
Yale University, Chemistry Gopher	gopher://yaleinfo.yale.edu:7700/11/OtherYaleGophers/chem

Index

A

Abitibi Environmental Technologies, 3, 195
ABS (*see* Aerated Biological Surfaces)
Achievable Emission Rate Clearinghouse, 199
Acid Rain Hotline (*see entry under* Environmental Protection Agency)
ACS (*see* American Chemical Society)
Aderco Fuel Additives, 3, 190
Advance Scientific and Chemical, Inc., 3, 187
Advanced Chemical Design, 4, 190
Advanced Chemicals, 4, 190
Advanced Visual Systems, 4, 187
Aerated Biological Surfaces, Inc. (ABS), 5, 203
Aerojet Chemicals, 5, 190
AET (*see* Alliance for Environmental Technology)
Agency for Toxic Substances and Disease Registry (ATSDR), 5, 186, 205
Agency for Toxic Substances and Disease Registry (ATSDR) Science Corner, 6, 205
AGRA Earth and Environmental Limited, 6, 200
AIChE (*see* American Institute of Chemical Engineers Web)
Air Risk Information Support Center Hotline (*see entry under* Environmental Protection Agency)
Ajax Chemicals, 6, 190
Akromold Inc., 7, 210
Akron University, 234
Akzo Nobel, 7, 190
Alberta Research Council (ARC), 7, 200
Alitea USA, 8, 181
Aliweb, 8, 213
Alliance for Environmental Technology (AET), 8, 195
AlliedSignal, 8, 209
Alpha Analytical Labs, 9, 182, 195
Alta Vista, 9, 213
Alternative Treatment Technology Information Center (*see entry under* Environmental Protection Agency)
ALToptronic AB, 9, 181
Aluminium Industry World Wide Web Server, 9, 187
American Chemical Society (ACS), 10, 186, 277
American Chemicals Company, Inc., 10, 187
American Institute of Chemical Engineers Web (AIChE), 10, 186
American Mold & Engineering, 11, 210
American Physical Society (APS), 11, 186
American Public Works Association (APWA), 11, 197
American Turbine Pump Co, 11, 195
American Vacuum Society (AVS), 12, 186
American Water Works Association (AWWA), 12, 196
Amoco Corporation, 12, 190
Analytical Chemistry and Instrumentation, 13, 182
Analytical Chemistry and Chemometrics Index, 13, 182
Analytical Chemistry Basics, 13, 182
Analytical Chemistry Hypermedia, 13, 182
Analytical Service Laboratories (ASL), 14
Analyticon Instruments Corporation, 14, 187
Applied Coatings & Linings, Inc., 14, 187
APS (*see* American Physical Society)
APWA (*see* American Public Works Association)
ARC (*see* Alberta Research Council)
ARCO Chemical, 14, 190
Argus Chemicals, 15, 190
Arizona State University, 219
ARL Home Page, 248
Armour College of Engineering, 221
ARSoftware's Online Internet Catalog, 15, 183
ASB Industries, 15, 210
Asbestos Ombudsman Clearinghouse/Hotline (*see entry under* Environmental Protection Agency)

ASD, 16, 181
ASEE Clearinghouse for Engineering Education, 16, 186
ASL (*see* Analytical Service Laboratories)
Aslchem International Inc, 16, 187
Aspen Technology, 16, 183
AspenTech Info E-Mail Discussion Lists, 253
Associated Rubber Company, 16, 209
ATI Cahn Company, 17, 210
ATI Orion, 17, 182
Atlantis Plastics, Inc., 17, 210
ATSDR (*see* Agency for Toxic Substances and Disease Registry)
Auburn University, 219, 234
Augias Environmental Corp., 17, 200
Ausimont USA, Inc., 18, 190
Australian Defense, 277
Automotive Recycling Mailing List, 18, 213
AVS (*see* American Vacuum Society)
AWWA (*see* American Water Works Association)

B

Baltzer Science Publishers, 18, 187
Basel Convention, 19, 206
BASF Corporation, 19, 190
Beckman Instruments, Inc., 19, 181
Beilstein Information Systems, 20, 187
Bethany College, 234
Bifurcation and Nonlinear Instability Laboratory, 20, 187
Bio Control Network, 20, 200
Bio-Online, 21, 184
Bioinorganic Chemistry Server, 241
BioSupplyNet, 21, 183
Black & Veatch, 21, 204
Boston College, 234
Boston University, 234
Boulder Scientific Co., 22, 190
Bourns, The Marlan and Rosemary, College of Engineering, 227
Brigham Young University, 219, 234, 248
Brookhaven National Laboratory, 234
Brown University, 219, 234
BUBL Information Service, 22
BUBL WWW Subject Tree -Chemical Engineering, 184
Buckman Laboratories, 22, 190
Bucknell University, 219
Burton Hamner's List of Internet Environmental Sources, 22, 206
Butler University, 234

C

CAD Centre at Strathclyde University, 184
Calgon Corporation, 23, 190, 203
California Institute of Technology, 219
California State University, Fresno, 234
California State University, Long Beach, 219
California State University, Sacramento, 219
California State University, Stanislaus, 234
Caltech, 234
Calvin College, 234
CambridgeSoft, Corp., 23, 183
CAPD (*see* Computer-Aided Process Design Consortium)
Carbolabs, Inc., 23, 190
Carbon Dioxide Information Analysis Center (CDIAC), 24, 205
Carnegie Mellon University, 219, 235, 248
CAS (*see* Chemical Abstract Service)
Case Western Reserve University, 220, 235
Catalytica, Inc., 24, 181
CCA , 184
CDC (*see* Center for Disease Control)
CDIAC (*see* Carbon Dioxide Information Analysis Center)
Celgene Corp., 24, 190
Center for Clean Technology, 206, 248
Center for Computational Quantum Chemistry, 241
Center for Disease Control (CDC) and Prevention, 25, 205, 277
Center for Molecular Design, 247
Center on Polymer Interfaces and Macromolecular Assemblies, 25
Central Instrument Facility, 235
Centre College, 235
CER (*see* Chemical Education Resources)
Ceramics and Industrial Minerals Home Page, 25, 187

CFD Resources Online, 26
CH2M Hill, 26, 204
Challenge, Inc., 26, 190
CHEMCOM, 254
CHEMCONF, 254
ChemConnect, 27
CHEMCORD, 254
CHEME-L, 255
CHEMED-L , 255
Chemical Abstract Service (CAS), 27, 184
  WWW Server, 187
Chemical Analytical Instrumentation
  Manufacturers, 181
Chemical Analytical Methods, Services, and
  Information, 182
Chemical Bulletin Board System , 255
Chemical Computer Software Analysis, 183
Chemical Concepts, 27, 184
Chemical Education Resources (CER), 28,
  187
Chemical Engineering, 214
Chemical Engineering: Information Indexes,
  28, 214
Chemical Engineering Meetings and
  Conferences, 186
Chemical Engineering Newsletter, 256
  Chemical Engineering: Professional
  Organization, 28, 214
Chemical Engineering Professional
  Organizations, 186
Chemical Engineering Programs, 219
Chemical Engineering: Research Organization
  and Laboratories, 29, 214
Chemical Engineering Sites all over the
  World, 28, 184
Chemical Industry Institute of Toxicology,
  235
Chemical Information Service, 184
Chemical Marketing Online (CHEMON), 29,
  188, 190
Chemical Physics Preprint Database, 29, 184,
  235
Chemical Process Modeling and Flowsheet
  Synthesis, 29, 188
Chemical Professional and Trading
  Organizations, 186
Chemical Services, 187
Chemical Supplier/Manufacturer, 190
Chemical Week Magazine, 30, 184
Chemie-Index (*see* Chemistry Index)
CHEMINF-L, 256
ChemInnovation Software, 30
ChemInnovation Software, 183
CHEMISTRY, 256
Chemistry, 30, 214
Chemistry Index /Chemie-Index (FU Berlin),
  30, 184
Chemistry Hypermedia Project, 247
Chemistry on the Internet, 31, 184
Chemistry Programs, 234
Chemistry Principles, 245
Chemistry Textbook Archive, 244
Chemistry Sites at Commercial Organizations,
  31, 214
Chemistry Sites at Nonprofit Organizations,
  31, 214
Chemistry Sites at Other Information
  Resources, 31, 214
ChemKey Database, 32, 184
CHEMLAB-L, 256
Chemnet, 190
CHEMON (*see* Chemical Marketing Online)
ChemSearch (Chemical Recycling), 32, 213
ChemSOLVE, 32, 185
ChemSource, 188
Chemsyn Science Lab., 33, 191
Chemtec Publishing, 33, 188
Chevron, 33, 191
Christian Brothers University, 220
Chromophore, Inc, 34, 191
Chrysler Corp. Recycling, 34, 213
Chugoku Kogyo Co., Ltd, 34, 191
Ciba-Geigy AG Basel, 34, 191
CICOURSE, 257
City College of the City University of New
  York, 220
City University of New York, Hunter College,
  235
Citylink, 35, 195
Civil Engineering, 35, 214
Clarkson University, 220
Clean Lakes Clearinghouse (*see entry under*
  Environmental Protection Agency)

Clean-Up Information Bulletin Board System (*see entry under* Environmental Protection Agency)
Clearinghouses and Hotlines (*see entry under* Environmental Protection Agency)
Clemson University, 220
Cleveland State University, 220
CLI International, Inc., 35, 188
Clorox Company, 35, 191
Coatings Industry Alliance, 36, 186
Colorado School of Mines, 220
Colorado State University, 220, 235
Columbia University, 220, 235
Communications for a Sustainable Future, 36, 206
Communicopia Environmental Research and Communications, 36, 200
Computer-Aided Process Design Consortium (CAPD), 36, 187
Connecticut College, 235
Consortium on Green Design and Manufacturing, 37, 206
Control Technology Center (*see entry under* Environmental Protection Agency)
Cooper Union, 220, 235
Cornell Theory Center, 277
Cornell University, 221, 235
CS Distribuidora, S.A. de C.V., 37, 188
CSC Chemistry Gopher in Finland, 277
CTD, Inc, 37, 191
CUNY (*see* City University of New York)
Custom Plastic Extrusions, 37, 210

D

Dalton Chemical Laboratories, Inc., 38, 191
Dartmouth College, 221
Daylight Chemical Information Systems, Inc., 38, 183
Deepwater Iodides, Inc., 38, 191
Department of Commerce (DOC), 39, 205
Department of Energy (DOE), 39, 205
Department of Energy, Office of Industrial Technologies (OIT), 39, 188
Department of Health and Human Services (DHHS), 40
Department of Labor (DOL), 40, 205
Design for the Environment, 41, 207
DHHS (*see* Department of Health and Human Services)
Diaz Chemcial Corp., 41, 191
Dielectric Polymers Inc., 41, 191, 209
DMP Corporation, 42, 203
DOC (*see* Department of Commerce)
DOE (*see* Department of Energy)
Dojindo Laboratories, 42, 191
DOL (*see* Department of Labor)
Dow Speciality Chemicals, 42, 191
Drago Group, 243
Drexel University, 221
Duke University, 235, 277
DuPont, 43, 191

E

EAM, 195
E & D Plastics, 43, 210
East Bay Municipal Utility District (EBMUD), 43
East Tennessee State University, 236
Eastern Minerals and Chemicals, 44, 188
Eastman Chemical Company, 44, 191
ECDM Internet Mailing List, 257
ECDM Group, 44, 207
ECI (*see* Environmental Concerns, Inc)
Eco-Glass Group, 44, 213
Ecocycle Newsletter, 45, 207
EFI (*see* Energy Federation)
EINET Galaxy, 45, 213
Electrochemical Society, Inc., 45, 186
Electronic Selected Current Aerospace Notices (E-SCAN), 46, 188
Eli Lilly and Company, 46, 191
Eller, Robert, Associates, 212
Elsevier Science B. V., 46, 188
EMAX Solution Partners, 46, 201
Emergency Planning and Community Right-to-Know Information Hotline (*see entry under* Environmental Protection Agency)
Emerson, Cherry L., Center for Scientific Computation, 236
Emission Factor Clearinghouse (*see entry under* Environmental Protection Agency)

Emory University, 236
Energy & Environmental Research Center, 47, 195
Energy and Environmentally Conscious Manufacturing, 47, 207
Energy Federation, Inc. (EFI), 47, 195
Engineered Rubber Products, 48, 209
Engineering Foundation, 48, 186
Enke Research Laboratory, 245
EnviroLink Network, 48, 195
Enviromine, 49, 195
Environment and Ecology, 49, 277
Environment Canada, 49, 207
Environment One Corporation (E/ONE), 49, 201
Environmental Concerns, Inc. (ECI), 50, 207
Environmental Engineering, 50, 214
Environmental Engineering Institutions, 248
Environmental Equipment Manufacturer/Supplier (Hardware), 195
Environmental Financing Information Network (*see entry under* Environmental Protection Agency)
Environmental Industry Web Site, 50, 202, 207
Environmental Information Service, 195
Environmental Library, 51
Environmental Library Environmental Professional's Guide to the Net (EPGN), 195
Environmental News Network, 51, 201
Environmental Professional and Trading Organization, 197
Environmental Professional's Guide to the Net (EPGN), 51
Environmental Protection Agency (EPA), 52, 198, 205
  Acid Rain Hotline, 53, 198
  Administration and Resources Management, 257
  Air and Radiation, 257
  Air Risk Information Support Center Hotline, 54, 198
  Alternative Treatment Technology Information Center, 54, 198
  Asbestos Ombudsman Clearinghouse/Hotline, 55, 198

Environmental Protection Agency—cont'd
Clean Lakes Clearinghouse, 55, 198
  Clean-Up Information Bulletin Board System, 56, 198
  Clearinghouses and Hotlines, 57, 198
  Control Technology Center, 58, 198
  Emergency Planning and Community Right-to-Know Information Hotline, 59, 198
  Emission Factor Clearinghouse, 60, 198
  Environmental Financing nformation Network, 61, 198
  Green Lights Program, 62, 198
  Hazardous Waste Ombudsman Program, 63, 198
  Indoor Air Quality Information Clearinghouse, 63, 198
  INFOTERRA, 64, 198
  Institute, 65, 198
  International Cleaner Production Information Clearinghouse, 66, 198
  Methods Information Communications Exchange, 67, 198
  Model Clearinghouse, 68, 198
  National Air and Radiation Environmental Laboratory (NAREL), 257
  National Air Toxics Information Clearinghouse, 69, 198
  National Exposure Research Lab (NERL), 257, 258
  National Health & Environmental Effects Research Lab (NHEERL), 258
  National Lead Information Center Hotline, 69, 198
  National Pesticide Information Retrieval System, 69, 198
  National Pesticide Telecommunications Network, 70, 198
  National Radon Hotline, 70, 199
  National Response Center, 71, 199
  National Risk Management Research Lab (NRMRL), 258
  National Small Flows Clearinghouse, 71, 199
  National Vehicle and Fuel Emissions Laboratory, 258
  Nonpoint Source Information Exchange, 72

Environmental Protection Agency—cont'd
Office of Air Quality Planning & Standards Technology Transfer Network Bulletin Board, 73, 199
Office of Radiation & Indoor Air-Las Vegas Laboratory, 259
Office of Research and Development Electronic Bulletin Board System, 73, 199
Office of Water Resource Center, 74, 199
OzonAction, 75, 199
Policy Planning and Evaluation, 259
Pollution Prevention Information Exchange System (PIES), 77, 199
Pollution Prevention Information Clearinghouse (PPIC), 76, 199
Radiation Surveillance Laboratory, 259
Reasonably Available Control Technology, Best Available Control Technology, and Lowest Achievable Emission Rate Clearinghouse, 78, 199
Regional Offices, 259
Research and Development, 261
Resource Conservation and Recovery Act/Superfund/Underground Storage Tank Hotline, 79, 199
Risk Communication Hotline, 79, 200
Robert S. Kerr Environmental Research Lab, 261
Safe Drinking Water Hotline, 80, 200
Solid Waste Assistance Program, 81, 200
Solid Waste and Emergency Response, 261
Stratospheric Ozone Information Hotline, 82, 200
33/50 Program, Special Projects Office (SPO), Office of Pollution Prevention and Toxics, 83, 200
Toxic Release Inventory UserSupport, 84, 200
Toxic Substances Control Act Assistance Information Service, 85, 200
Wastewater Treatment Information Exchange, 85, 200
Wastewi$e Program, 86, 207
WWW Server, 196
Environmental Recycling Hotline, 86, 213
Environmental Sensors, 86, 195
Environmental Services, 200
Environmental Technologies USA, Inc., 87, 213
Environmental Wastewater Information Service , 202
Environmental Wastewater Treatment
Equipment Manufacturer/Supplier, 203
Environmental Wastewater Treatment Facilities (Municipal), 203
Environmental Wastewater Treatment Services, 204
Environmentally Conscious Design and Manufacturing Lab, 88, 207
Enviroene, 87, 207
EnviroWeb, 196, 207
E/ONE (*see* Environment One Corporation)
EPA (*see* Environmental Protection Agency)
EPA-Water, 262
EPGN (*see* Environmental Professional's Guide to the Net)
E-SCAN (*see* Electronic Selected Current Aerospace Notices)
ETS International, Inc., 88, 201
Evaluation of Artificial Wetlands, 88, 203
Exchange, The, 88, 213
Exfluor Research Corp., 191

F

Falcon Software, 89, 183
Farrell Research, 89, 201
Fast Heat, 90, 210
FDA (*see* Food and Drug Administration)
Federal Printing Office, 205
FedWorld Information Network, 90, 205
Fiberglass World, 90, 191
Finishing Industry Home Page, 91, 188
Fisher Scientific, 91, 191
Florida Agriculture and Mechanical University, 221
Florida Institute of Technology, 221
Florida State University, 236
FMC Corporation, 91, 191
Food and Drug Administration (FDA), 92, 205
Ford Motor Co. Recycling, 92, 213
Foundation for Cross-Connection Control and Hydraulic Research, 92, 203

Francis, John B., College of Engineering, 229
FRC International, 93, 213
Friends of Earth Home Page, 93, 196
Furuuchi Chemical Co., 93, 192

G

Garden State Laboratories, 94, 182
Gas Processors Association (GPA), 94, 186
Gaussian Basis Set Order Form, 239
GE (*see* General Electric)
Gelman Sciences, 94, 192
GENBBB (*see* Generic Bulletin Board Builder)
Genentech, Inc., 95, 192
General Electric (GE) Plastics, 95, 209
General Electric WWW Server, 95, 192
General Plastex, 96, 211
Generic Bulletin Board Builder (GENBBB), 96, 185
George Washington University, 236
Georgetown University, 236
Georgia Institute of Technology, 221, 236
Georgia Tech, 248
GEOSYN, 262
Geraghty & Miller, Inc., 96, 201
Gilchrist Polymer Center, 97, 211
Gilson, Inc., 97, 181, 188
Global Network of Environment and Technology (GNET), 97, 201
Global Recycling Network, Inc., 98, 213
GNET (*see* Global Network of Environment and Technology)
Goodyear Tire & Rubber Company, 98, 192
Government Printing Office (GPO), 98
GPA (*see* Gas Processors Association)
GPO (*see* Government Printing Office)
Green Design Initiative, 99, 207
Green Engineering, 99, 215
Green Lights Program (*see entry under* Environmental Protection Agency)
Green Market, 99, 201
Greenfield Environmental, 100, 201
GreenSoft Corporation, 100, 201
Greenspan Technology, 100, 204
Guide to Chemical Engineering, 100, 185

H

H&S Chemical Co., 101, 192
HAAKE, 101, 209
Hahnemann University, 236
Hamline University, 236
Hampford Research, Inc., 101, 192
Hampshire Chemical Corp., 102, 192
Hampton University, 221
Harbec Plastics, 102, 211
Harvard University, 236
Harvey Mudd College of Engineering and Science, 221
Hazardous Materials Management, 102, 202
Hazardous Waste Ombudsman Program (*see entry under* Environmental Protection Agency)
Hewlett-Packard Analytical, 1031, 181
HIRIS-L, 262
Hitachi Instruments, Inc., 103, 181
Howard University, 221
Huls America Inc., 103, 192
Husky Injection Molding Systems, 103, 211
Hydrocomp, Inc., 104, 201
Hydromantis, 104, 204
Hypercube, Inc., 104, 183

I

IAQ (*see* Indoor Air Quality)
IBM World Wide Web, 105, 183
ICI Fiberite, 105, 209
ICS-L, 262
IDES (*see* Integrated Design Engineering Systems)
Idetec, S.A. de C.V., 105, 192
IGUPT (*see* Institute for Gas Utilization and Processing Technologies)
Illinois Institute of Technology, 221
Illinois State University, 236
Imperial College, 277
Imperiali Lab, 105
Indiana University, 236, 277
Indigo Instruments, 106, 181
Indofine Chemical Co., 106, 192

Indoor Air Quality (IAQ) Publications, 106, 201
Indoor Air Quality Information Clearinghouse (*see entry under* Environmental Protection Agency)
Industrial Assessment Center, University of Florida, 207
Industrial Plastics and Paints, 107, 209
Industrial Services International, 107, 209
IndustryLink, 107, 188
IndustryNET, 108, 196, 202
Info-Labview Mailing List, 108, 182, 263
Information Technology for Environmentally Conscious Design, Construction, and Manufacturing, 108, 207
INFOTERRA (*see entry under* Environmental Protection Agency)
Ingot Metal Company, Ltd., 108, 195
Inktomi, 109
Institute for Gas Utilization and Processing Technologies (IGUPT), 109, 188
Institute of Paper Science and Technology, 222
Integrated Design Engineering Systems (IDES), 109, 211
Integrating Environment and Development, 109, 207
Intera Inc., 110, 201
Interactive Simulations, Inc., 110, 188
Interchem Corporation, 110, 192
Interduct, 110, 207
International Cleaner Production Information Clearinghouse (*see entry under* Environmental Protection Agency)
International Organization for Standardization (ISO), 111, 112, 207
International Society of Heterocyclic Chemistry, 111, 186
Internet Chemistry Resources, 111, 112, 185
Introduction to the SAFETY mailing list, 263
Iowa General Chemistry Network, 236
Iowa State University, 222, 237
ISO (*see* International Organization for Standardization)
ISO 14000 Information, 112, 207
IVAM Environmental Research, 113, 208

J

J.M. Huber Corporation, 113, 192
Jamestown Tooling & Machining, 113, 211
Jandel Scientific, 114, 183
JEOL USA, Inc., 114, 188
Johns Hopkins University, 222, 277
Jost Chemical, 114
Journal of Chemical Physics Express Service, 277
Journal of Biological Chemistry, 240

K

K.R. Anderson Co., Inc, 115, 192
Kansas State University, 222
Keith Ceramic Materials, 115, 192
Khem Products, Inc., 115, 185
Kimberlyte Inc., 116, 189
Knight-Ridder Information - Dialog, 116, 185

L

Laboratory Equipment Exchange, 116, 186
Laboratory for Magnetic Resonance and NMR Periodic Table, 241
Lackie & Associates, 117, 213
Lafayette College, 222
Lakewood Systems, 117, 195
Lamar University, 222
Lancaster Synthesis, Inc., 117, 192
Lanxide Coated Products, 118, 192
Lauren Manufacturing, 118, 211
LCA at the University of Toronto, 118, 206
Lebanon Valley College, 237
Leco Corporation, 119, 181
Lehigh University, 222, 237
Library of Congress (LOC), 119, 205, 277
Life Cycle Analysis, 206
LOC (*see* Library of Congress)
Loras College, 237
Los Alamos National Laboratory, 237
Louisiana Tech University, 222, 237
Low Gravity Transport Phenomena Laboratory , 119, 189

Loyola Marymount University, 237
Loyola University Chicago, 237

M

Macquarie University, 277
MacroFAQS, 119, 211
Maddox, Robert N., Chemical Engineering Technical Reference Center, 227
Manhattan College, 222
Manufacturing, 108
Manufacturing, Inktomi, 213
Marshall University, 237
Massachusetts Institute of Technology (M.I.T.), 222, 237
MAT-DSGN, 263
Material Safety Data Sheets (MSDS), 277
Materials and Electrochemical Research (MER) Corporation, 120, 193
Materials System Laboratory, The, 120, 208
Maxima Plastics, 120, 211
McCormick School of Engineering and Applied Science, 224
McNeese State University, 222
MDL Informations Systems, Inc., 121, 183
Medical Branch in Galveston (*see* University of Texas)
Melamine Chemicals, Inc., 121, 193
Membrane Technology Group, 121, 202
MER (*see* Materials and Electrochemical Research)
Merck & Company, Inc., 121, 193
Merkert Chemistry Department, 234
Metcalfe Plastics Corporation, 122, 211
Methanex, 122, 193
Methods Information Communications Exchange (*see entry under* Environmental Protection Agency)
Michigan State University, 223, 237, 278
Michigan Technological University, 223, 237
Microanalytics Instrumentation, 122, 181
MicroMath Scientific Software, Inc., 123, 183
Micromeritics Instrument Corp., 123, 181
Middle Tennessee State University, 237
Millipore Corporation, 123, 193
Misco International, Inc., 124, 193
Mississippi State University, 223, 238
Missouri Western State College, 238
Miton Products, 124, 193
M.I.T. (*see* Massachusetts Institute of Technology)
Mittelhauser, 124, 201
Mobil Corporation, 124, 193
Model Clearinghouse (*see entry under* Environmental Protection Agency)
Moldflow Ltd, 125, 211
Molecular Modeling Home Page, 238
Molecular Modeling Laboratory, 247
Molecular Simulations, Inc., 125, 183
Molten Metal Technology, 125, 195
Monsanto Company, 126, 193
Montana State University-Bozeman, 223
Montreal Protocol on Substances that Deplete the Ozone Layer, 126, 208
Morflex, Inc., 126, 193
Morris Environmental, 126, 201
Mother Lode Plastics, 127, 211
Multibase, 127, 211

N

NAEP (*see* National Association of Environmental Professionals)
Nanotechnology Papers, 240
National Air Toxics Information Clearinghouse (*see entry under* Environmental Protection Agency)
National Association of Environmental Professionals (NAEP), 127, 197
National Center for Supercomputing Applications, 238
National Environmental Information Service, 186
National Filter Media Corp., 128, 203
National Institute of Environmental Health Sciences (NIEHS), 128, 196, 205, 278
National Institutes of Health (NIH), 129, 205, 238, 278
National Institute of Occupational Safety and Health (NIOSH), 129, 205
National Institute of Standards and Technology (NIST), 129, 206, 238

National Key Centre for Design, 130, 208
National Lead Information Center Hotline (*see entry under* Environmental Protection Agency)
National Library of Medicine (NLM), 130, 206, 278
National Oceanic and Atmospheric Administration (NOAA), 131, 206, 278
National Pesticide Information Retrieval System (*see entry under* Environmental Protection Agency)
National Pesticide Telecommunications Network (*see entry under* Environmental Protection Agency)
National Pollution Prevention Center for Higher Education, 131, 208
National Radon Hotline (*see entry under* Environmental Protection Agency)
National Response Center (*see entry under* Environmental Protection Agency)
National Science Foundation, 132
National Small Flows Clearinghouse (*see entry under* Environmental Protection Agency)
National Technical Information Service (NTIS), 132, 206
National Technology Transfer Center (NTTC), 133, 196
NCSA ChemViz Group, 238
NCSA Digital Information System, 238
NDB (*see* Nucleic Acid Database)
Nelson, Leonard C., College of Engineering, 233
Nerken, Albert, School of Engineering, 220
Nest Group, Inc., 133, 181
Neste Resins North America, 133, 193
Nevada Technical Associates, Inc., 134, 189
New Jersey Institute of Technology, 223
New Mexico Institute of Mining and Technology, 248
New Mexico State University, 223
New Traditions Cuurriculum Project, 247
New York University Medical Center, 238
Nicolet Instruments, 134, 181
NIEHS (*see* National Institute of Environmental Health)
NIH (*see* National Institutes of Health)
NIOSH (*see* National Institute of Occupational Safety and Health)
NIST (*see* National Institute of Standards and Technology)
NLM (*see* National Library of Medicine)
NOAA (*see* National Oceanic and Atmospheric Administration)
Nonpoint Source Information Exchange (*see entry under* Environmental Protection Agency)
Norquay Technology Inc., 134, 193
North American Catalysis Society, 135, 187
North Carolina Agricultural and Technical State University, 223
North Carolina State University, 223, 238, 278
Northeastern University, 223
Northern Illinois University, 238, 278
Northwestern University, 224, 278
NSF International, 135, 182
NTIS (*see* National Technical Information Service)
NTTC (*see* National Technology Transfer Center)
Nucleic Acid Database (NDB) Archive, 135, 185, 278
Nutting Environmental of Florida, 136, 201

O

O2 Global Network, 136, 208
Oberlin College, 238
Occupational Safety and Health Administration (OSHA), 136, 206
Office of Air Quality Planning & Standards Technology Transfer Network Bulletin Board System (*see entry under* Environmental Protection Agency)
Office of Industrial Productivity and Energy Assessment, 136, 208
Office of Industrial Technologies (OIT), 39, 188
Office of Pollution Prevention and Compliance Assistance, 137, 208

Office of Research and Development Electronic Bulletin Board System (*see entry under* Environmental Protection Agency)
Office of Water Resource Center (*see entry under* Environmental Protection Agency)
Ohio State University, 224, 239
Ohio Supercomputer Center, 239
Ohio University, 224, 239
Ohio Valley Plastics Partnership, 137, 211
OIT (*see* Department of Energy, Office of Industrial Technologies)
Oklahoma State University, 224, 239
Old Line Plastics, 137, 211
OM Group, Inc. (OMG), 138, 193
Online Databases, Libraries and Facilities, 138, 185
Oregon Graduate Institute of Science & Technology, 239
Oregon Graduate Institute of University of Cincinnati, 248
Oregon State University, 224, 239
OSHA (*see* Occupational Safety and Health Administration)
Oxford Molecular Group, 138, 183
OzonAction (*see entry under* Environmental Protection Agency)

P

P2 Information Service, 206
Pacific Northwest Laboratories, 239
Pacific Lutheran University, 239
PAD, 263
Palm International, 138, 193
Papros Inc., 139, 193
Park Equipment Company, 139, 203
Park Scientific Instruments, 139, 210
Parr-Green, 140, 211
Penn State University, 224, 239, 248, 278
Performance Plastics, 140, 211
Perkin-Elmer Corporation, 140, 181
Pfizer International, 141, 193
Pharm-Eco Laboratories, Inc, 141, 193
Phoenix Polymers, Inc., 141, 210
Pilot Chemical Company, 142, 193
PIXE Analytical Laboratories, Inc., 142, 182
Plasti Dip Plastic Spray, 142, 211
Plastic Bag Association, 142, 213
Plastic Express, 142, 211
Plastic Technology Center, 143, 212
Plastics Group, The, 143, 212
Plastics Hotline, 143, 212
Plastics Network, 144, 209
PLI, 209
PLM, 209
PLS, 210
Polaris Plastic Sales, 144, 212
Pollution Prevention and Waste Minimization, 144, 208
Pollution Prevention Information Clearinghouse (PPIC) (*see entry under* Environmental Protection Agency)
Pollution Prevention Program Database, 144, 208, 278
PolyLinks, 145, 209
Polymer Science Research Center, 245
PolyNet, 145, 187
PolySort, 145, 209
Polytechnic University (Brooklyn), 224
Pomona College, 239
Powder Page, 145, 185
Prairie View A&M University College of Engineering and Architecture, 224
Pressure Chemical Co., 146, 194
Princeton University, 224, 239
Pro-Mold, 146, 212
Progressive Products, Inc., 146, 189
Publishers, 146, 196
Purdue University, 224, 236, 240

Q

QCPE Server, 278
QMR Plastics, 147, 212
Quadrax Corp., 147, 210
Quality Chemicals Inc., 147, 194
Questral/Orbit, 147
Quimica Carnot, S.A., 148, 194

R

Radiation Research Journal, 148, 189
Reasonably Available Control Technology, Best Available Control Technology, and Lowest Achievable Emission Rate Clearinghouse (*see entry under* Environmental Protection Agency)
Recycler's World, 148, 213
Recycling, 213
Recycling WWW Servers, 148, 213
Regensburg Theoretical, 278
Reid Engineering, 149, 204
Reilly Industries, Inc., 149, 194
Remco Engineering, 149, 203
Rensselaer Polytechnic Institute, 185, 225, 240
Replas, 149, 212
Rescan International Inc., 150, 201
Research Organics, 150, 194
Resource Conservation and Recovery Act/Superfund/Underground Storage Tank Hotline (*see entry under* Environmental Protection Agency)
Resource Development Associates, 201
Reuther Mold & Machine, 150, 212
Reverse Logistics (Take-Back Technologies), 151, 208
Rhone-Poulenc Surfactants and Specialties, 151, 194
Rhyme Industries, 151, 201
Ribbon-Jet Tek, 151, 213
Rice University, 225, 240, 278
Ridout Plastics, 152, 212
Risk Communication Hotline (*see entry under* Environmental Protection Agency)
Ritrama Group, 152
RJF International Corporation, 152, 210
Ro-Mai Industries, 153, 212
Robert Eller Associates, 153, 212
Rockefeller University, 240
Rohm and Haas Company, 153, 210
Rose-Hulman Institute of Technology, 225
Rothberg, Tamburini and Winsor, Inc., 153, 204
Russ College of Engineering and Technology, 224
Rutgers University, 225, 240

S

Sachem, Inc., 154, 194
SAF Bulk Chemicals, 154, 189
Safe Drinking Water Hotline (*see entry under* Environmental Protection Agency)
Sajar Plastics, 154, 212
Sam Houston State University, 240
Sampling Systems by PMMI, Inc., 155, 181
San Jose State University, 225
SCC Environmental, 155, 201
Schlumberger , 194
Schroinger, Inc., 155, 183
Scientific and Technical Information Network, 212
Scientific Instrument Services (SIS), 156, 181
Scott Butner's list of Internet Environmental Sources, 156, 208
Scripps Research Institute, 240
SCS Engineers, 156, 204
SensonCorp Ltd., 157, 194
ServicesNational Science Foundation , 206
Sewage in Puget Sound, 157, 203, 279
Shell Chemical Company, 157, 194
Showa Chemical's Database, 158, 189
Sigma Chemical Company, 158, 194
SimaPro3, 158, 208
SIS (*see* Scientific Instrument Services)
Smith, L.C., College of Engineering and Computer Science, 226
Smith's Enterprises, 158, 212
Smith's Scientific Services, 159
SoftShell Online, 183
Solid Waste Assistance Program (*see entry under* Environmental Protection Agency)
Solstice, 159, 208
Solutions Software, 159, 202
Some Chemistry Resources on the Internet, 185
South Dakota School of Mines and Technology, 225
Southern Illinois University at Carbondale, 240
Specialty Recycling Services, 160, 213
Spectrocell, 160, 182

Spirex, 160, 212
SRI International, 161, 189
Stanford University, 225, 240, 248
State Technical Institute of Memphis, 225
State University of New York at Albany, 240
State University of New York at Binghamton, 241
State University of New York at Buffalo, 225, 241
State University of New York at Stony Brook, 241
Stevens Institute of Technology, 225
STN International, 161, 162, 185, 212
Stormceptor, Inc., 162, 202
Stratospheric Ozone Information Hotline (*see entry under* Environmental Protection Agency)
Struktol Company of America, 162, 210
Sumitomo Chemical Co., 163, 194
Surface-l, 264
Sustainable Development Online, 163, 208
Synthetech, 163, 194
Syracuse University, 226
Systems Realization Laboratory, 164, 208

T

Take-Back Technologies (*see* Reverse Logistics)
TCC (*see* The Thermochemical Calculator)
Tech Mold, 164, 212
TechExpo (TM), 202
Technology and Trade Inc., 165, 212
Technology, Business and the Environment Program, The, 164, 209
TeleChem International, Inc., 165, 189
Texaco OnLine, 165, 194
Texas A&M University, 226, 241, 248
Texas Christian University (TCU), 241
Texas Tech University, 226, 241
Thayer School of Engineering, 221, 249
Thermochemical Calculator, The, (TCC), 165, 189
33/50 Program, Special Projects Office (SPO), Office of Pollution Prevention and Toxics (*see entry under* Environmental Protection Agency)
Thomas, 166, 185
Thomas Register Of American Manufacturers, 166, 185
Titlenet, 279
Toagosei Co., Ltd., 166, 194
Toxic Release Inventory User Support (*see entry under* Environmental Protection Agency)
Toxic Substances Control Act Assistance Information Service (*see entry under* Environmental Protection Agency)
TRAC (*see* Trends in Analytical Chemistry)
Transportation Resources, 166, 189
TreeEco, 167, 213
Trends in Analytical Chemistry (TRAC), 167, 182
Tres English's Sustainable Development Pages, 167, 209
Tripos, Inc., 167, 183
Trojan Technologies Inc., 168, 203
TSD Central (Hazardous Waste Central), 168, 202
Tufts University, 226, 241
Tulane University, 226

U

UCLA CCT, 168, 209
Ultrasonic Testing OnLine Journal, 169, 182
Umetrics, 169, 194
UNICAMP, IQ Resources, 279
Unique Tire Recycling, 169, 195
Universal Enviromental Technologies, 170, 203
Universal Plastics, 170, 212
University of Akron, 226
University of Alabama, Birmingham, 241, 242
University of Alabama, Huntsville, 226, 241, 242
University of Arizona, 226
University of Arkansas, 227
University of California, Berkeley, 227, 242
University of California, Davis, 227, 242, 279
University of California, Irvine, 227, 242
University of California, Los Angeles, 227, 242

University of California, Riverside, 227, 242
University of California, San Francisco, 242
University of California, San Diego, 227, 242
University of California, Santa Barbara, 227, 242, 279
University of California, Santa Cruz, 242
University of Chicago, 242
University of Cincinnati, 227, 242
University of Colorado, 228
University of Connecticut, Storrs, 228, 242
University of Dayton, 228
University of Delaware, 228, 243
University of Denver, 243
University of Detroit, Mercy, 228
University of Florida, 228, 243, 249
University of Georgia, 241
University of Hawaii, 243
University of Houston, 228, 243, 279
University of Idaho, 228, 243
University of Illinois, Chicago, 228, 243
University of Illinois, Urbana-Champaign, 228, 243, 279
University of Iowa, 229
University of Kansas, 229
University of Kentucky, 229, 243
University of Louisville Speed Scientific School, 229
University of Maine, 229
University of Maryland, Baltimore County, 229, 243
University of Maryland, College Park, 243
University of Massachusetts, Amherst, 229, 243
University of Massachusetts, Lowell, 229, 244
University of Memphis, 244
University of Miami, 244
University of Michigan, 229
University of Minnesota, 229, 244
University of Mississippi, 230, 244
University of Missouri-Columbia, 230
University of Missouri-Rolla, 230, 244
University of Missouri-St. Louis, 244, 279
University of Nebraska, Kearney, 244
University of Nebraska, Lincoln, 230, 244
University of Nevada, Reno, 230, 244
University of New Hampshire, Durham, 230
University of New Haven, 230
University of New Mexico, 230, 244, 245
University of New Orleans, 244
University of North Carolina, 245
University of North Dakota, 230
University of Notre Dame, 230, 241
University of Oklahoma, 231, 245
University of Oregon, 245
University of Pennsylvania, 231, 245
University of Pittsburgh, 231, 245
University of Puerto Rico, 231
University of Rhode Island, 231, 245
University of Rochester, 231, 245
University of South Alabama, 231
University of South Carolina, 231
University of South Dakota, 245
University of South Florida, 231, 245
University of Southern California, 231, 245
University of Southern Colorado, 246
University of Southern Mississippi, 245, 246
University of Southwestern Louisiana, 231
University of Tennessee, Chattanooga, 232
University of Tennessee, Knoxville, 232, 246, 279
University of Texas at Arlington, 246
University of Texas at Austin, 232, 246
University of Texas, Medical Branch in Galveston, 246
University of Toledo, 232
University of Tulsa, 232
University of Utah, 232, 246, 279
University of Virginia, 232
University of Washington, 232, 246
University of Wisconsin-Madison, 232, 247, 279
University of Wisconsin-Oshkosh, 247
University of Wyoming, 232, 247
U.S. Technology, 168, 210
Utah Water Research Laboratory, 170, 204
Utility, Power, Wastewater Plant's E-mail Directory, 171, 203

Van Dorn Demag, 171, 210
Vanderbilt University, 233
VanLare, Frank E., Wastewater Treatment Facility, 93, 203

Varian Analytical Instruments, 171, 182
VERONICA, 171, 279
Villanova University, 233, 247
Virginia Polytechnic Institute and State University, 233
Virginia Tech, 247
Viscona Limited, 172, 194
Vista Chemical Company, 172, 194
VPI, 172, 212

W

Washington and Lee University, 279
Washington State University, 233
Washington University in St. Louis, 233, 247
Wastewater Engineering, 172, 215
Wastewater Information, 173, 215
Wastewater Treatment Information Exchange (*see entry under* Environmental Protection Agency)
Wastewater Solutions, 173, 204
Wastewater Web, 173, 202
Wastewi$e Program (*see entry under* Environmental Protection Agency)
Water & Wastewater Utilities, 173, 203
WaterOnLine, 174, 202
Waters Corporation, 174, 182
Water Resources Management, Inc., 173, 204
WAU Process Engineering, 175, 196
Waterweb, 175, 202
Waterworld, 175, 202
Wavefunction, Inc., 175, 184
Wayne State University, 233
West Virginia Institute of Technology, 233
West Virginia University, 233, 247
Western Michigan University, 247
Whiting School of Engineering, 222
Widener University, 247
Wilkes University, 247
Wiltec Research Company, Inc., 176, 183
WindowChem Software, Inc., 176, 184
Witt Plastics, 176, 212
WMX Technologies, 177, 202
Worcester Polytechnic Institute, 233, 247
World Wide Chemnet, Inc., 177, 189
World Wide Web Engine, 213
World Wide Web Virtual Library, 214
World-Wide Web Virtual Library: Environmental Engineering, 196
World-Wide Web Virtual Library: Wastewater Engineering, 196
WWTNET, 177, 202
WWW Chemicals, 178, 189

Y

Yale University, 233, 248, 279
Youngstown State University, 233

Z

Zefon Manufacturing, 178, 210
Zipperling Kessler & Co, 178, 210